Distributed Event-triggered Control

Zhongkui Li • Bin Cheng • Weihao Song •
Shiqi Zhang

Distributed Event-triggered Control

Scalability and Robustness

Springer

Zhongkui Li
Department of Mechanics and Engineering Science
Peking University
Beijing, China

Bin Cheng
Department of Control Science and Engineering
Tongji University
Shanghai, China

Weihao Song
Department of Mechanics and Engineering Science
Peking University
Beijing, China

Shiqi Zhang
Department of Mechanics and Engineering Science
Peking University
Beijing, China

ISBN 978-981-99-8169-4 ISBN 978-981-99-8170-0 (eBook)
https://doi.org/10.1007/978-981-99-8170-0

This Springer imprint is published by the registered company Springer Nature Singapore Pte Ltd.
The registered company address is: 152 Beach Road, #21-01/04 Gateway East, Singapore 189721, Singapore

Paper in this product is recyclable.

Preface

The past two decades have witnessed an explosion of theoretical research with respect to cooperative control of multi-agent systems, due primarily to its immense application prospects in both civilian and military realms. Representative application scenarios include, but are not limited to, forest-fire prevention, unmanned logistics, satellite formation, aircraft rendezvous, and salvo attack of multi-missile system. For a multi-agent system, the inter-agent communication via local information sharing among neighboring agents without doubt plays a vital role in implementing collaborative tasks. However, frequent inter-agent communication will inevitably consume a large number of network bandwidth resources that are usually constrained in practice. To reduce the communication frequency and hence save the limited network resources, distributed event-triggered control schemes, which selectively trigger the inter-agent communication according to certain predefined rules, have drawn considerable research attentions in recent years.

The objective of this book is to summarize the authors' original and systematic contributions on distributed event-triggered control of multi-agent systems, with a particular emphasis on the *scalability* and *robustness* issues. The *scalability* herein means that the distributed control schemes are independent of any global information on the network (e.g., the eigenvalues of the Laplacian matrix) and scalable with respect to the network size. On the basis of the related research overview in Chap. 1 and mathematical preliminaries in Chap. 2, the fully distributed and scalable event-triggered consensus problem is investigated in Chap. 3 for general linear multi-agent networks, where the dependence of any global information of the network is avoided and only event-triggered communication is needed. Chapter 4 continues to consider the problem of fully distributed consensus with discrete communication and control, in which discrete communication among neighboring agents, discrete control updating of each agent, and independence of global information are simultaneously taken into account. Chapter 5 continues

The original version of the book has been revised. A correction to this book can be found at https://doi.org/10.1007/978-981-99-8170-0_8

to study fully distributed event-triggered affine formation maneuver control, for which both numerical simulations and hardware experiments are provided. For these proposed fully distributed adaptive event-triggered protocols, the *scalability* is certainly guaranteed since they rely on neither continuous communications nor global information of the network.

Robustness is a fundamental requirement in control theory. Any implementable control laws, event-triggered schemes of course included, should maintain certain *robustness* with respect to various uncertainties and disturbances. Especially, the event-triggering mechanisms used in the literature are expressed in the time domain. There is a natural gap between the ubiquitous frequency-domain uncertainties (e.g., unmodelled dynamics) and the time-domain event triggering, which exactly explains why corresponding results are really rare in the existing literature. This book endeavors to shed new light on the design of robust event-triggered algorithms with a view to accommodating either frequency-domain or matching uncertainties. Specifically, Chap. 6 considers the distributed consensus and tracking control problems for multi-agent systems subject to matching uncertainties as well as the cooperative output regulation problem by designing fully distributed adaptive event-triggered controllers. Chapter 7 analyzes the robustness of the event-triggered consensus algorithms against frequency-domain uncertainties and provides a novel operator-theoretic approach to bridge time-domain event triggering and frequency-domain uncertainties.

The authors are grateful to the great efforts and encouragements of our colleagues, collaborators, and students, who have helped in many ways in making this book possible. In particular, we would like to express our sincere thanks to Professor Lin Huang, Professor Zhisheng Duan, Professor Jie Chen (Tongji University), Professor Bin He, Professor Jie Chen (CityU of Hong Kong), Professor Guanrong Chen, Professor Gang Feng, Professor Zhengtao Ding, Professor Wei Ren, Professor Lihua Xie, Professor Xiang Chen, and many other people for their enduring support and kind help. The first author would also like to thank his current/former students, including Dr Yuezu Lv, Zeze Chang, and Junjie Wang for their contributions to parts of the results in this book.

Finally, we would like to acknowledge the financial support from the National Key R&D Program of China (2022ZD0116401, 2022ZD0116400), the National Natural Science Foundation of China under grants U2241214, T2121002, 61973006, 62373008, 62103302, and 62203016, the Beijing Natural Science Foundation of China under grant JQ20025, and the Shanghai Sailing Program under grant 21YF1450300.

Beijing, China — Zhongkui Li
Shanghai, China — Bin Cheng
Beijing, China — Weihao Song
Beijing, China — Shiqi Zhang
September 2023

Contents

Chapter 1
Introduction

1.1 Research Background

With the technological advances of lightweight and miniaturized computation, sensing and communication units, the past two decades have seen a surge of interest in multi-agent systems due to their diverse applications in industries ranging from transportation, agriculture to astronomy [50, 53, 77, 84]. The research of multi-agent systems is initially motivated by various biological phenomena in nature, such as the coordinated foraging behavior of ant colony, the formation flying of geese, the fish schooling and migration, and the huddling of emperor penguins. It has been well recognized that a large number of individual organisms are capable of exhibiting some sophisticated collective behaviors via the group interaction and information sharing. Accordingly, in parallel to the gradually deep understanding of natural collective behaviors, considerable research attention has been devoted to investigating how to coordinate the behaviors of multi-agent systems and facilitate the emergence of swarm intelligence in recent years.

Typically speaking, a multi-agent system (e.g., unmanned aerial vehicle (UAV) swarm, automobile system, aerospace system, and vehicle platooning) is composed of a large number of geographically distributed agents, and each agent has certain ability to perceive the external environment via sensors, generate the customized command via controllers, and execute some physical actions via actuators. By means of group cooperation, it is possible to conduct some complicated or dangerous tasks that cannot be solely finished by an individual. For example, a group of small and inexpensive UAVs can be deployed in a perilous environment to execute certain complex tasks, such as the cooperative sensing and surveillance, large-scale information relay, and the like, while a big and expensive UAV may fail in these scenarios. Compared with single systems, multi-agent systems with inter-agent cooperation are able to offer numerous advantages, such as improved operational effectiveness, reduced costs, higher flexibility, and increased robustness against external attacks [84].

Z. Li et al., *Distributed Event-triggered Control*,
https://doi.org/10.1007/978-981-99-8170-0_1

In response to the above-identified advantages, cooperative control of multi-agent systems, whose primary purpose is to drive the agents to achieve a desirable collective behavior, has rapidly become an active research area in both academia and industry. Since the limited network resource and communication range usually constitute a major concern in practical applications, an important problem with respect to cooperative control of multi-agent systems is how to design distributed control schemes by only utilizing the local information exchanged among neighboring agents. Towards such a direction, a wide range of interesting and practical topics have emerged in recent years, which includes consensus, formation, tracking, flocking, rendezvous, coverage control, and distributed optimization. Among these topics, the consensus problem, initially developed in the realm of computer science, has become the most fundamental one in cooperative control, as well as the key to solving other coordination problems. Consensus control refers to the design of a distributed control strategy for each agent according to task requirements and performance indicators, such that some states of interest reach an agreement through interactions among neighboring agents [84, 121]. The corresponding local control strategy that needs to be designed is often called consensus protocol or consensus law. Detailed discussions and extensions on this issue can be found in [53, 83].

It should be pointed out that in practical applications of multi-agent systems, the number of agents is usually large, unknown, or even time-varying (due to the changes of mission environment), and the sensing or communication ability of each single agent is often limited. Such issues have posed significant challenges to the design of distributed control schemes with practical feasibility. To this end, a critical approach is to guarantee that in the designed cooperative control laws, each agent only uses the information (e.g., local states and outputs) of itself and its neighbors, and does not rely on the global structure or any information related to the overall network.

1.2 Introduction of Event-Triggered Control

The rapid development of modern communication technology has made it feasible to wirelessly connect different plants or components deployed at different places, which gives rise to the prompt popularity of networked control systems. However, for most of practical networked systems, an ineluctable issue is that the available bandwidth of the communication network and the power sources are usually limited. Traditional control strategies that require continuous or periodic communication can be inefficient in terms of resource usage. To overcome this drawback, the event-triggered control has emerged as a promising alternative. In event-triggered control, communication is triggered only when certain signals in the system exceed a designed threshold, rather than being continuously executed. This approach optimizes the use of communication resources and has been extensively studied in recent years.

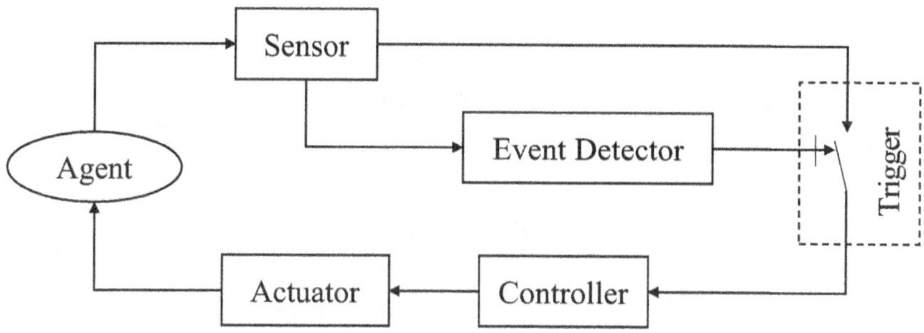

Fig. 1.1 Event-triggered control structure of single system

Event-triggered control is first proposed for a single system to trigger a data transmission from sensors to controllers and actuators only if a well-defined error signal exceeds a given bound as shown in Fig. 1.1; see the references [2, 34, 35, 37, 90, 117]. As an essential part of most diverse applications, small embedded units are required to execute a variety of tasks including the relay of information packets in multi-hop communication schemes, monitoring physical quantities, and computation. For example, physically distributed sensor/actuator networks are responsible for collecting and processing information, and reacting to such information through actuation. However, microprocessors often have limited resources. In this sense, it is important to determine when and/or how to transmit information and execute control tasks.

To make the idea of event-triggered control more clear, we take an example of a system [90], whose dynamics satisfy

$$\dot{x} = Ax + Bu, \; x \in \mathbf{R}^n, \; u \in \mathbf{R}^r, \tag{1.1}$$

with a feedback controller

$$u = k(\hat{x}), \tag{1.2}$$

where $\hat{x}$ is the estimated state. Denote $e = \hat{x} - x$ as the measurement error. The closed-loop dynamics are given as

$$\dot{x} = Ax + Bk(\hat{x}). \tag{1.3}$$

Then, the closed-loop system is asymptotically stable if $|e(t)| \leq \sigma|x(t)|$, where $0 < \sigma < 1$. We can set the triggering instant as

$$t_{k+1} \triangleq \inf\{t > t_k \mid |e(t)| \geq \sigma|x(t)|\}, \tag{1.4}$$

with $t_0 \triangleq 0$. Event-based control offers some clear advantages with respect to the traditional methods, such as saving energy and minimizing the number of control actions, when handling practical constraints [35].

1.3 Multi-Agent Event-Triggered Control

As mentioned earlier, multi-agent systems, being a special class of networked systems, have received considerable research attention in recent years [22, 32, 33, 53, 55, 74, 80, 101]. It is well known that the information exchanges or communications among neighboring agents and across the network are necessary to achieve the desired cooperative and collective behaviors of multi-agent systems. Traditional control strategies require communication process being carried out at each time for continuous-time algorithms or at all iterations for discrete-time algorithms. However, the bandwidth of the communication network and the power source of the agents are inevitably constrained in practical engineering. That is to say, in networked control applications, it makes sense to only transmit information when certain signal in the system is larger than a threshold value [37], as depicted in Fig. 1.2. In such a context, the multi-agent event-triggered control strategy is developed as an important means for avoiding continuous communications.

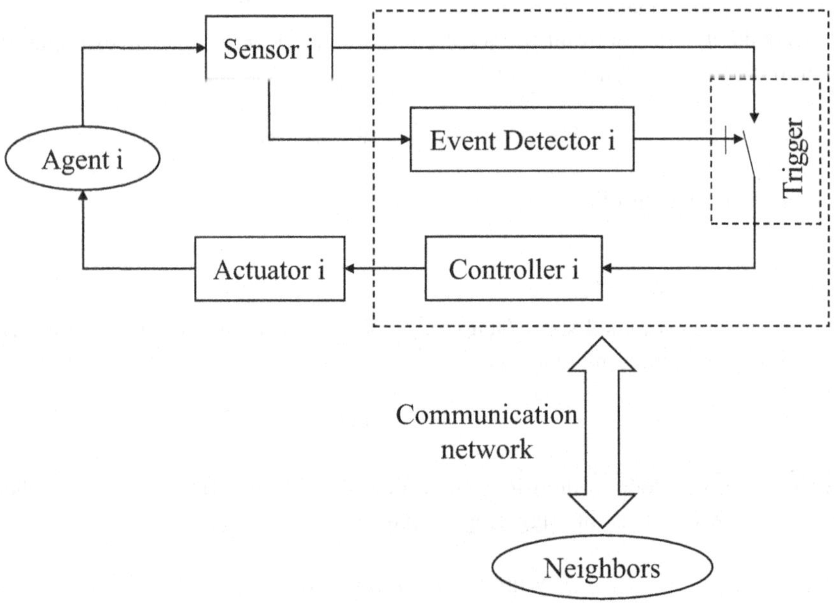

Fig. 1.2 Event-triggered control structure of multi-agent systems

It should be pointed out that the multi-agent event-triggered control problem poses many new challenges, even though event-based control laws are similar to classic continuous-time protocols in form. Generally, most of the challenges are caused by a fact that only sampled information rather than real-time relative information between neighboring agents can be utilized. Moreover, the well-known separation principle generally does not hold for event-triggered control networks. Note that the most critical task for users is to co-design both triggering functions and control laws to guarantee the achievement of control objective and the exclusion of undesirable Zeno behavior. It is vital to making trade-off between the control performance and the communication cost by designing appropriate event-based protocols and choosing proper control parameters. A comprehensive survey on the event-based interaction and control of multi-agent systems was presented in [74]. Generally speaking, the research attention on multi-agent event-triggered control can be classified into four main issues: dynamics, topology, triggering condition, and control objective.

Up to now, a variety of event-triggered control schemes have been developed for multi-agent systems with different dynamics. For example, in [21, 69], event-triggered and self-triggered consensus algorithms were proposed for single-integrator agents over undirected connected topologies. Decentralized event-triggered consensus algorithms were presented in [87] for both single- and double-integrator multi-agent systems. In [69, 87], novel event-triggered conditions were proposed, which do not require continuous communication to achieve consensus. Researchers also proposed event-based consensus strategies for multi-agent systems with general linear models in [31, 102, 112, 119]. The authors of [56] studied the average consensus problem of discrete-time multi-agent systems with limited data transmission rates. Observer-based output feedback event-triggered protocols were designed in [113]. Additionally, the authors of [38] studied the event-based cooperative output regulation problem of heterogeneous networks based on homogeneous dynamics.

It is also important to consider the existence of uncertainties and disturbances in multi-agent systems, as they widely exist in various kinds of dynamics and systems. Several works have studied event-based control for multi-agent systems in the presence of uncertainties and disturbancec. For instance, the work [11] studied consensus disturbance rejection problem with event-based communications and proposed distributed protocols for multiple neutrally stable agents and general linear agents. The work [97] considered event-driven adaptive control for uncertain nonlinear systems and designed three different strategies. In [62], the authors studied fixed-time event-based consensus for multi-agent systems suffering from nonlinear uncertainties. The authors of [114] studied event-based adaptive fuzzy tracking problem of multi-input and multi-output nonlinear uncertain agents. In [6], the authors designed distributed dynamic event-triggered protocols for networked Euler-Lagrange systems with parametric uncertainties.

Regarding the topology of multi-agent systems, the aforementioned works mainly considered the case of undirected topology where the Laplacian matrix is symmetric. However, in many practical cases, the topologies may be directed or switching and may not satisfy the connectivity condition. Thus, several works have proposed event-triggered algorithms for directed and switching network topologies. For instance, the works [30, 85] proposed distributed event-triggered algorithms for directed network topologies. The work [120] studied consensus of multi-agent systems with a time-varying topology and proposed an event-based dynamic feedback scheme. In [1], the authors proposed an event-driven protocol for networks with switching communication graphs. In [17], the authors proposed decentralized event-based controllers for leader-follower networks under fixed or switching graphs.

The existing works on event-triggered control of multi-agent systems have proposed various triggering conditions to address the issue of excessive communication among agents. Note that the triggering conditions are often only necessary but not sufficient conditions. From the perspective of thresholds, the triggering conditions can be classified into three types: state-based threshold, time-based threshold, and hybrid threshold that combines the former two types. In terms of the dynamics of triggering conditions, they can be static or dynamic. In addition, the triggering conditions can be node-based or edge-based functions. Most of the existing works on event-triggered control have focused on node-based triggering conditions. Actually, edge-based triggering conditions have also been proposed in some studies. Specifically, edge-based triggering functions are designed for each edge, and if they are triggered, only the two nodes connected by the edge need to communicate. The authors of [7, 94, 96] studied sampled-data consensus of multi-agent systems using edge-based event-triggered communications. Edge-based event-triggered output regulation of heterogeneous multi-agent networks was studied in [71]. Edge-based event-based control problem for network systems was addressed in [95] via integral quadratic constraints. Other triggering conditions have also been proposed, which include triggering conditions with periodic checking intervals, stochastic triggering conditions, and self-triggering conditions. The choice of appropriate triggering condition depends on the specific needs of the user, and the trade-off between control performance and interaction frequency.

On the other hand, numerous studies have explored the event-triggered control for a variety of objectives, such as leaderless consensus, coordinated tracking, output feedback consensus, and robust consensus in the presence of external disturbances. For example, researchers in [16, 98, 107] investigated the event-triggered leader-follower consensus problem for multivariable multi-agent systems with one leader. In [91], event-based algorithms were designed to track trajectory for nonlinear multi-agent systems with bounded reference trajectory and bounded exogenous input. The authors in [16] studied distributed event-based tracking of multivariable multi-agent systems, while the authors in [107] proposed distributed event-based algorithms to solve the tracking-based flocking problem. In [64], distributed adaptive output feedback protocols were designed for linear multi-agent systems under general directed graphs with a dynamic leader. Additionally, the distributed event-based sliding-mode controllers were proposed in [104] to ensure tracking performance for uncertain multi-agent systems, and the adaptive event-based fuzzy tracking protocols were developed in [114] to guarantee transient performance for multi-agent systems with nonlinear uncertainties.

1.4 Scalability and Robustness Issues

Note that in multi-agent systems, it is important to consider scalability when designing distributed protocols. Scalability refers to that the designed protocols are scalable and should be independent of the scale of the network. Going back to the case of continuous-time communications, the design parameters of controllers often rely on the eigenvalues of the Laplacian matrix associated with the network topology. To this end, some distributed procedures via local interactions among neighboring agents were proposed in [24, 100] to estimate the eigenvalues of the Laplacian matrix. However, when running these distributed estimation procedures simultaneously with consensus protocols, the convergence and feasibility of the overall algorithm remain unclear and questionable. To deal with such a problem, the work [55] replaced the constant parameter by time-varying parameters and proposed scalable adaptive protocols.

When considering the event-triggered communications in multi-agent systems, we also need to propose scalable protocols. As pointed out in [87, 102, 119], certain global information of the network, in terms of nonzero eigenvalues of the Laplacian matrix associated with the communication graph, is generally required in the aforementioned papers to determine parameters in either the control laws or the triggering functions. Therefore, the event-based consensus protocols in the aforementioned works are actually not fully distributed. Even though similar issues also exist in the cooperative control of multi-agent systems via continuous-time communications, like [55], it is quite different and more challenging to design fully distributed event-triggered protocols. Specifically, the constraint of event-based communications results in structured nonlinear networks with asynchronous and aperiodic control and communication signals. How to design fully distributed and scalable event-triggered consensus protocols for general linear or more complex multi-agent networks is quite challenging.

On the other hand, the plant to be controlled may be subject to various kinds of uncertainties. Typically, when we use lower-order simplified models to represent the actual complicated models, or represent nonlinear or even unknown ones with their linearized/identified versions, we have to design controllers with certain robustness guarantee in the sense of stability and performance for a whole set of plants within which the actual models may vary. This problem is typically handled via the $\mathcal{H}_\infty$ and structure singular value theories [118]. Compared to the single-agent case, the incorporation of network topology in the multi-agent distributed control settings renders the robustness problem a more inclusive one, including the robustness against plant dynamics and topology perturbations. For example, the authors of [92] studied the robust consensus problem under additive and coprime-factor uncertainties, respectively. The work [51] studied the robust controller design problem against the network perturbations.

It is well known that distributed control algorithms with continuous and ideal communications should be robust to various kinds of uncertainties. Then, a natural question arises that whether distributed event-triggered algorithms also possess the robustness against these kinds of uncertainties. To answer this question, it is necessary to adopt the frequency-domain robust control tools, such as the small gain theorem and the μ analysis. However, essentially speaking, the event-triggered algorithms belong to a special branch of the aperiodic sampling algorithms, whose definition, modeling, and methodology are all based on the time-domain analysis. More specifically, the triggering function, which decides whether the agent updates its state estimation and broadcasts it to its neighbors, is expressed in a time-domain form. Moreover, it characterizes a time-domain pointwise inequality constraint of the sampling error. Therefore, the analysis and design of the event-triggered control problem have almost always been based on the Lyapunov stability analysis, which is severely different from the robust analysis and synthesis tools mentioned above. Because of the systematic gap between time-domain event triggering and the frequency-domain uncertainties, the robustness of event-triggered control of multi-agent systems against frequency-domain uncertainties still remains a very challenging problem.

Motivated by the above discussions, this book focuses on the scalability and robustness issues in distributed event-triggered control of multi-agent systems. Specifically, this book proposes fully distributed adaptive event-triggered protocols, relying on neither continuous communications nor global information of the whole network, and therefore being scalable with respect to the network scale and the communication topology. Moreover, in order to deal with widespread uncertainties, this book proposes novel robust event-triggered control algorithms, and especially presents an operator-theoretic approach, which bridges the gap between the time-domain event triggering and the frequency-domain uncertainties. This book well solves the above two core issues of distributed event-triggered control, and hence is expected to be of great importance to the control and related communities.

1.5 Overview of This Book

This book provides a concise yet systematic study of the distributed event-triggered control problems of multi-agent systems with general linear, uncertain, and typical nonlinear agent dynamics. Important topics, including consensus, tracking with a dynamic leader, formation control, and robustness, are systematically considered.

This monograph consists of seven chapters. A brief introduction of these chapters is given as follows.

Chapter 1 introduces the background of cooperative control of multi-agent systems and reviews the development history of event-triggered control.

Chapter 2 provides some basic mathematical preliminaries on the graph theory, the robust control theory, and the stability theory.

Chapter 3 considers the scalable event-triggered consensus problem for general linear multi-agent networks. To avoid relying on global information of the network graph, two types of fully distributed adaptive event-based protocols are proposed, which introduce adaptive gains into both the control laws and the triggering functions using the local state information sampled at event instants.

Chapter 4 continues to consider the problem of fully distributed consensus with discrete communication and control. Fully distributed event-triggered protocols are established, simultaneously possessing three features, namely, discrete communications among neighboring agents, discrete control updating of each agent, and independence of any global information of the network.

Chapter 5 considers the affine formation maneuver control problem for linear multi-agent systems with event-triggered communication over directed graphs. The in-degree information of the graph is incorporated to produce fully distributed event-triggered affine formation control laws. Numerical simulations and hardware experiments are provided to demonstrate the effectiveness of the proposed schemes.

Chapter 6 studies the robust control problem of multi-agent systems with time-domain uncertainties. Fully distributed robust adaptive event-triggered controllers are designed to achieve consensus and tracking control for linear multi-agent systems subject to matching uncertainties. Cooperative output regulation of heterogeneous linear multi-agent networks is also considered.

Chapter 7 investigates the robustness of event-triggered consensus algorithms against frequency-domain uncertainties. A novel operator-theoretic perspective is provided to characterize the event-triggered mechanisms, by showing that the sampling errors are images of finite-gain $\mathcal{L}_2$ stable operators acting on the consensus error of the sampled states. Using this operator-theoretic approach, the robustness and robust performance of event-triggered consensus algorithms in the presence of frequency-domain uncertainties are revealed.

Chapter 2
Mathematical Preliminaries

This chapter introduces some mathematical preliminaries that will be utilized in the sequel of this book. Basic notations are first provided, and then some results in the graph theory are presented. Subsequently, certain definitions and fundamental results in the areas of stability and robustness are recalled.

2.1 Notations

The notations used throughout this book are fairly standard. $\mathbf{R}^{n\times m}$ denotes the linear space of all $n \times m$-dimensional matrices and $\mathbf{R}^n$ stands for the n-dimensional Euclidean vector space. I_N represents the N-dimensional identity matrix and 1_N the N-dimensional vector whose elements are all equal to 1. $\mathbf{1}$ denotes a column vector with all entries equal to 1. $\text{diag}\{a_1, \cdots, a_n\}$ denotes the diagonal matrix whose diagonal elements are equal to $a_1, \cdots, a_n$. For a symmetric matrix A, $\lambda_{\max}(A)$ and $\lambda_{\min}(A)$ denote the maximum and minimal eigenvalues of A, respectively. For a vector $x \in \mathbf{R}^n$, $\|x\|$ denotes its 2-norm. $\otimes$ stands for the Kronecker product. The set of all real rational stable transfer matrices is denoted by $\mathcal{RH}_\infty$. The H_∞ norm of a transfer function $G(s)$ in $\mathcal{RH}_\infty$ is denoted by $\|G(s)\|_\infty = \sup_{Res\geq 0} \bar{\sigma}(G(s))$. The set of all real rational strictly proper transfer functions is denoted by $\mathcal{RH}_2$. For any $G(s) \in \mathcal{RH}_2$, its H_2 norm is denoted by $\|G(s)\|_2 = \dfrac{1}{2\pi} \int_{-\infty}^{\infty} G(-j\omega)G(j\omega)d\omega$.

2.2 Graph Theory

The information flow among the agents is usually described by a graph $\mathcal{G} = (\mathcal{V}, \mathcal{E})$, where $\mathcal{V} = \{1, 2, \cdots, N\}$ denotes the set of vertices (also called nodes) and $\mathcal{E} \subseteq \mathcal{V} \times \mathcal{V}$ represents the set of edges. If the edge $(i, j) \in \mathcal{E}$, then node j can receive

Z. Li et al., *Distributed Event-triggered Control*,
https://doi.org/10.1007/978-981-99-8170-0_2

information from node i, and node i is called a neighbor of node j. The set of neighbors of node i is denoted by $\mathcal{N}_i$. A graph $\mathcal{G}$ is said to be undirected if for any $(i, j) \in \mathcal{E}$, one has $(j, i) \in \mathcal{E}$. A directed graph is said to be strongly connected if there exists a directed path from every node to every other node. A undirected graph $\mathcal{G}$ is connected if for any $i, j \in \mathcal{V}$, there exists an undirected path from node i to node j.

A (rooted) directed tree is a directed graph in which every node has exactly one parent except for one node, called the root, which has no parent and has directed paths to all other nodes. A directed tree is called spanning if it connects all the nodes in the graph, which implies that there is at least one root node connected with a path to all the other nodes. A graph is a leader-follower graph, if it has and only has one root node. Note that a directed tree is a leader-follower graph but the converse may not be true. A graph is said to contain a directed spanning tree if a subset of the edges forms a directed spanning tree, which is equivalent to that the graph has at least one node having directed paths to all other nodes.

Three types of matrices associated with graphs, namely, the adjacency matrix, the Laplacian matrix and the incidence matrix, are introduced as follows. The *adjacency* matrix of a graph $\mathcal{G}$ is denoted as $\mathcal{A} = [a_{ij}]_{N\times N}$, satisfying that $a_{ij} > 0$ if $(j, i) \in \mathcal{E}$ and $a_{ij} = 0$ otherwise, in which the nonzero a_{ij} is called the weight of edge (j, i). For an unweighted graph, a_{ij} is set equal to one if $(j, i) \in \mathcal{E}$. A common assumption is that self loops do not exist, i.e., $(i, i) \notin \mathcal{E}$ or $a_{ii} = 0$ for any $i \in \mathcal{V}$. Letting $d_i = \sum_{j=1}^{N} a_{ij}$ be the in-degree of node i and $H = \text{diag}\{d_1, \cdots, d_N\}$ be the degree matrix, the *Laplacian* matrix of graph $\mathcal{G}$ is defined as $\mathcal{L} = H - \mathcal{A}$. Define $D = [d_{ij}]_{N\times m}$ as the *incidence* matrix, satisfying that $d_{ij} = -1$ if node i is the tail of the jth edge, $d_{ij} = 1$ if node i is the head of the jth edge, and $d_{ij} = 0$ otherwise. It is easy to see that the sum of each column of D is equal to 0.

For illustration, an undirected connected graph and a strongly connected graph with six nodes are given in Fig. 2.1, where the lines with directional arrows denote directed edges and the lines without arrows denote undirected edges.

Based on the above definitions, the adjacency matrices, Laplacian matrices and incidence matrices of these two graphs can be written as follows:

$$\mathcal{A}_a = \begin{bmatrix} 0 & 1 & 0 & 1 & 0 & 1 \\ 1 & 0 & 1 & 0 & 0 & 0 \\ 0 & 1 & 0 & 0 & 0 & 0 \\ 1 & 0 & 0 & 0 & 1 & 0 \\ 0 & 0 & 0 & 1 & 0 & 1 \\ 1 & 0 & 0 & 0 & 1 & 0 \end{bmatrix}, \quad \mathcal{A}_b = \begin{bmatrix} 0 & 1 & 0 & 1 & 0 & 1 \\ 1 & 0 & 1 & 0 & 0 & 0 \\ 0 & 1 & 0 & 0 & 0 & 0 \\ 0 & 0 & 1 & 0 & 1 & 0 \\ 1 & 0 & 0 & 0 & 0 & 0 \\ 0 & 0 & 0 & 0 & 1 & 0 \end{bmatrix},$$

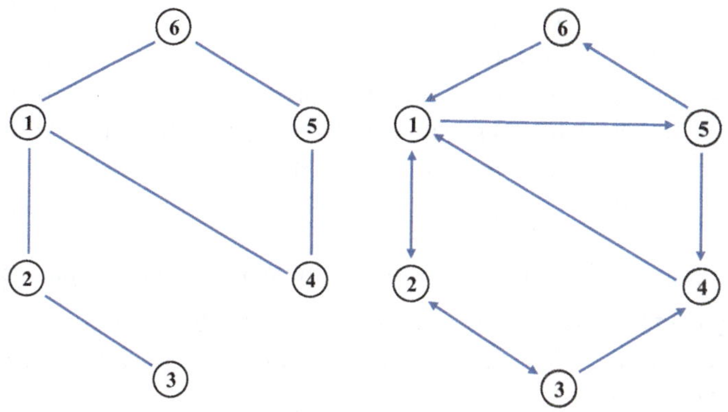

(a) (b)

Fig. 2.1 Different graphs with six nodes: (**a**) an undirected connected graph, (**b**) a strongly connected graph

$$\mathcal{L}_a = \begin{bmatrix} 3 & -1 & 0 & -1 & 0 & -1 \\ -1 & 2 & -1 & 0 & 0 & 0 \\ 0 & -1 & 1 & 0 & 0 & 0 \\ -1 & 0 & 0 & 2 & -1 & 0 \\ 0 & 0 & 0 & -1 & 2 & -1 \\ -1 & 0 & 0 & 0 & -1 & 2 \end{bmatrix},$$

$$\mathcal{L}_b = \begin{bmatrix} 3 & -1 & 0 & -1 & 0 & -1 \\ -1 & 2 & -1 & 0 & 0 & 0 \\ 0 & -1 & 1 & 0 & 0 & 0 \\ 0 & 0 & -1 & 2 & -1 & 0 \\ -1 & 0 & 0 & 0 & 1 & 0 \\ 0 & 0 & 0 & 0 & -1 & 1 \end{bmatrix},$$

$$D_a = \begin{bmatrix} -1 & 0 & -1 & -1 & 0 & 0 \\ 1 & -1 & 0 & 0 & 0 & 0 \\ 0 & 1 & 0 & 0 & 0 & 0 \\ 0 & 0 & 1 & 0 & -1 & 0 \\ 0 & 0 & 0 & 0 & 1 & -1 \\ 0 & 0 & 0 & 1 & 0 & 1 \end{bmatrix},$$

$$D_b = \begin{bmatrix} -1 & 1 & 1 & -1 & 1 & 0 & 0 & 0 & 0 & 0 \\ 1 & -1 & 0 & 0 & 0 & -1 & 1 & 0 & 0 & 0 \\ 0 & 0 & 0 & 0 & 0 & 1 & -1 & -1 & 0 & 0 \\ 0 & 0 & -1 & 0 & 0 & 0 & 0 & 1 & 1 & 0 \\ 0 & 0 & 0 & 1 & 0 & 0 & 0 & 0 & -1 & -1 \\ 0 & 0 & 0 & 0 & -1 & 0 & 0 & 0 & 0 & 1 \end{bmatrix}.$$

In what follows, we will provide some useful lemmas regarding the graphs.

Lemma 2.1 ([72]) *For an undirected graph $\mathcal{G}$, the Laplacian matrix $\mathcal{L} = DWD^T$, where $W = \text{diag}\{w_1, w_2, \cdots, w_m\}$ with w_m denoting the weight corresponding to the mth edge.*

Lemma 2.2 ([53]) *(i) Zero is an eigenvalue of $\mathcal{L}$ with $\mathbf{1}$ as a right eigenvector and all nonzero eigenvalues have positive real parts. (ii) For a directed graph, zero is a simple eigenvalue of $\mathcal{L}$ if and only if $\mathcal{G}$ has a directed spanning tree. (iii) For an undirected graph, zero is a simple eigenvalue of $\mathcal{L}$ if and only if $\mathcal{G}$ is connected. The smallest nonzero eigenvalue $\lambda_2(\mathcal{L})$ of $\mathcal{L}$ satisfies that $\lambda_2(\mathcal{L}) = \min_{x \neq 0, \mathbf{1}^T x = 0} \frac{x^T \mathcal{L} x}{x^T x}$.*

Lemma 2.3 ([68]) *For a strongly connected graph $\mathcal{G}$ with Laplacian matrix $\mathcal{L}$, the matrix $\hat{\mathcal{L}} = R\mathcal{L} + \mathcal{L}^T R$ represents a weighted symmetric Laplacian matrix of an undirected connected graph, where $R = \text{diag}\{r_1, \cdots, r_N\} > 0$ with $r = [r_1, \cdots, r_N]$ being the left unit eigenvector of $\mathcal{L}$ corresponding to the zero eigenvalue. Moreover, $\min_{z^T x = 0} x^T \hat{\mathcal{L}} x \geq \frac{\lambda_2(\hat{\mathcal{L}})}{N} x^T x$, where $\lambda_2(\hat{\mathcal{L}})$ is the smallest nonzero eigenvalue of $\hat{\mathcal{L}}$ and z is a vector with all positive elements.*

2.3 Robust Control, Stability, and Other Tools

In this section, some preliminary lemmas and definitions on the stability and robustness theory are summarized.

Lemma 2.4 (Barbalat's Lemma, [44]) *If x and $\dot{x}$ are bounded, and $\int_0^{+\infty} x^T(\tau) x(\tau) d\tau < +\infty$, then $x(t) \to 0$ as $t \to +\infty$.*

Lemma 2.5 (Young's Inequality, [4]) *If a and b are nonnegative real numbers and p and q are positive real numbers such that $\frac{1}{p} + \frac{1}{q} = 1$, then $ab \leq \frac{a^p}{p} + \frac{b^q}{q}$.*

Lemma 2.6 (Comparison Lemma, [47]) *Consider the scalar differential equation:*

$$\dot{u} = f(t, u), \ u(t_0) = u_0,$$

where $f(t, u)$ is continuous in t and locally Lipschitz in u, for all $t \geq 0$ and all $u \in J \subset R$. Let $[t_0, T)$ be the maximal interval of existence of the solution $u(t)$, *and suppose $u(t) \in J$ for all $t \in [t_0, T)$. Let $v(t)$ be a continuous function whose upper right hand derivative $D^+v(t)$ satisfies the differential inequality:*

$$D^+v(t) \leq f(t, v(t)), \ \ v(t_0) \leq u_0,$$

with $v(t) \in J$ for all $t \in [t_0, T)$. Then, $v(t) \leq u(t)$ for all $t \in [t_0, T)$.

Definition 2.1 ([20]) Let $\mathcal{H}_1$ and $\mathcal{H}_2$ be two Banach spaces. An operator $\phi(\cdot) : \mathcal{H}_1 \mapsto \mathcal{H}_2$ is called a linear operator if the following two conditions hold:

(1) $\phi(x + y) = \phi(x) + \phi(y)$, for $\forall x, y \in \mathcal{H}_1$,
(2) $\phi(\lambda x) = \lambda\phi(x)$, for $\forall\ \lambda \in \mathbf{R}$.

Definition 2.2 The $\mathcal{L}_2$ norm of a signal f is defined as

$$\|f\|_2 = \sqrt{\int_0^\infty f^*(t)f(t)dt} = \sqrt{\frac{1}{2\pi}\int_{-\infty}^\infty f^*(j\omega)f(j\omega)d\omega},$$

where $\mathcal{L}_2$ denotes the Banach space with $\mathcal{L}_2$ norm well defined.

Definition 2.3 ([20]) Letting $\phi(\cdot) : \mathcal{L}_2 \mapsto \mathcal{L}_2$ be an operator such that for $\forall x \in \mathcal{L}_2$, $\|\phi(x)\|_2 \leq \gamma\|x\|_2 + \beta$, where $\gamma > 0$ and $\beta > 0$ are positive constants, then this operator is called a finite-gain operator with operator norm $\|\phi\|_\infty \leq \gamma$.

Lemma 2.7 ([20]) *Let $\phi(\cdot) : \mathcal{L}_2 \mapsto \mathcal{L}_2$ denote a linear (finite-gain) operator in the time domain and suppose that $y(t) = \phi(x(t))$, where $y(t)$ and $x(t)$ are vectors in the $\mathcal{L}_2$ space. Denote by $y(s)$ and $x(s)$ the Laplace transformations of $y(t)$ and $x(t)$, respectively. Then, it follows that $y(s) = \Delta(x(s))$, where $\Delta(\cdot) : \mathcal{L}_2 \mapsto \mathcal{L}_2$ denotes a linear (finite-gain) operator in the frequency domain.*

Lemma 2.8 (Small Gain Theorem, [20]) *Supposing that $G(\cdot), \Delta(\cdot) : \mathcal{L}_2 \mapsto \mathcal{L}_2$ are finite-gain operators with operator norms $\|G\|_\infty = \gamma_1$ and $\|\Delta\|_\infty = \gamma_2$, the system interconnection shown in Fig. 2.2 is internally stable, if $\gamma_1\gamma_2 < 1$.*

Definition 2.4 ([118]) Let $\mathbf{\Delta}$ represent the set of structured finite-gain $\mathcal{L}_2$-stable operators. For $M \in \mathbf{R}^{m\times n}$, $\mu_{\mathbf{\Delta}}(M)$ is defined as

$$\mu_{\mathbf{\Delta}}(M) = \frac{1}{\min\{\bar{\sigma}(\Delta) : \Delta \in \mathbf{\Delta}, \det(I - M\Delta) = 0\}},$$

unless no $\Delta \in \mathbf{\Delta}$ makes $\det(I - M\Delta)$ singular, in which case $\mu_{\mathbf{\Delta}}(M) \triangleq 0$.

Lemma 2.9 ([118]) *Let $\mathbf{\Delta}$ represent the set of structured finite-gain $\mathcal{L}_2$-stable operators. The loop shown in Fig. 2.2 is well-posed and internally stable for all $\Delta \in \mathbf{\Delta}$ with operator norm $\|\Delta\|_\infty \leq \gamma$ if and only if $\sup_{\omega\in\mathbf{R}} \mu_{\mathbf{\Delta}}(G(j\omega)) < \frac{1}{\gamma}$, where $\mu_{\mathbf{\Delta}}(\cdot)$ denotes the structured singular value.*

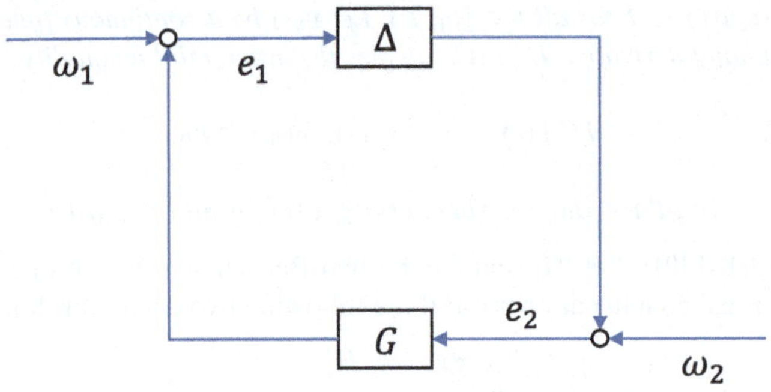

Fig. 2.2 System interconnection

Lemma 2.10 ([118]) *Assume that* $G(s) = \begin{bmatrix} G_{11}(s) & G_{12}(s) \\ G_{21}(s) & G_{22}(s) \end{bmatrix}$ *and* $\mathbf{\Delta} = \begin{bmatrix} \mathbf{\Delta}_1 & 0 \\ 0 & \mathbf{\Delta}_2 \end{bmatrix}$ *is the set of finite-gain* $\mathcal{L}_2$ *stable block diagonal operators with compatible dimensions with* $G(s)$. *Then,* $\mu_{\mathbf{\Delta}}(G(j\omega)) \leq \sqrt{\|G_{11}(j\omega)\|^2 + \|G_{22}(j\omega)\|^2 + 2\|G_{12}(j\omega)\|\|G_{21}(j\omega)\|}$, $\forall \omega \in \mathbf{R}$. *Moreover, suppose that* $\mathbf{\Delta} = \begin{bmatrix} \mathbf{\Delta}_1 & & \\ & \mathbf{\Delta}_2 & \\ & & \mathbf{\Delta}_3 \end{bmatrix}$ *and* $G = \begin{bmatrix} G_{11} & G_{12} & G_{13} \\ G_{21} & G_{22} & G_{23} \\ G_{31} & G_{32} & G_{33} \end{bmatrix}$. *Then,*

$$\begin{aligned} \mu_{\mathbf{\Delta}}^2(G(j\omega)) \leq \|G_{11}(j\omega)\|^2 + \|G_{22}(j\omega)\|^2 + \|G_{33}(j\omega)\|^2 \\ + 2\|G_{12}(j\omega)\|\|G_{21}(j\omega)\| + 2\|G_{13}(j\omega)\|\|G_{31}(j\omega)\| \\ + 2\|G_{23}(j\omega)\|\|G_{32}(j\omega)\| \end{aligned}$$

Lemma 2.11 ([118]) *Let* $\gamma > 0$ *and* $\mathcal{F}_u(M, \Delta) = M_{22} + M_{21}\Delta(I - M_{11}\Delta)^{-1}M_{12}$ *denote the upper linear fractional transformation with respect to* Δ. *For all* $\Delta \in \mathbf{\Delta}$ *with* $\|\Delta\|_\infty \leq \gamma$, *the transfer function* $\mathcal{F}_u(G_p, \Delta)$ *is internally stable and* $\|\mathcal{F}_u(G_p, \Delta)\|_\infty < \frac{1}{\gamma}$ *if and only if* $\sup_{\omega \in \mathbf{R}} \mu_{\mathbf{\Delta}_p}(G_p(j\omega)) < \frac{1}{\gamma}$, *where* $\mathbf{\Delta}_p = \begin{bmatrix} \mathbf{\Delta} & \\ & \mathbf{\Delta}_f \end{bmatrix}$.

2.4 Notes

The materials of Sect. 2.2 are mainly based on [53, 68, 72] and those of Sect. 2.3 are mainly adopted from [4, 20, 44, 47, 118].

Chapter 3
Fully Distributed Consensus Control with Event-Triggered Communication

The crucial task of the event-triggered (also called event-based) consensus problem is to design distributed event-based protocols or schemes, which consist of event-based control laws and event triggering functions, to ensure that the states of the agents reach agreement. The control laws rely on the local information sampled at discrete event instants and the triggering functions determine these event instants, at which time each agent broadcasts its state over the network or measures the relative states/outputs with respect to its neighbors. It should be noted that these event-triggered consensus protocols in the existing literature, e.g., [16, 17, 28, 102, 113], require the knowledge of global eigenvalue information of the network graph, and thereby are not truly distributed. This chapter devotes to removing this limitation and presenting fully distributed and scalable event-triggered consensus schemes.

In Sect. 3.1, we formulate the event-triggered consensus control problem, the objective of which is to design distributed event-triggered protocols such that consensus can be achieved and the Zeno behavior can be excluded. In Sect. 3.2, we first propose a static event-triggered consensus protocol for general linear multi-agent systems to achieve consensus with discrete communications. This static event-triggered protocol, as expected, similarly as those in [16, 17, 28, 102, 113], depends on global eigenvalue information of the network. Motivated by this drawback, we then propose a distributed node-based adaptive event-based protocol in Sect. 3.3. One distinct feature of the proposed adaptive event-based protocol is that both the control laws and the triggering functions include time-varying coupling weights, which can compensate the lack of global eigenvalue information of the graph. We show that the proposed protocol can guarantee consensus, and meanwhile rule out the Zeno behavior at any finite time by showing that the interval between two arbitrary triggering instants is lower bounded by a strictly positive value.

In Sect. 3.4, we further propose another type of edge-based adaptive event-triggered consensus protocol, which, different from the node-based protocols proposed in Sect. 3.3 where triggering functions are designed for each node, implements local triggering functions for each edge of the network topology to

Z. Li et al., *Distributed Event-triggered Control*,
https://doi.org/10.1007/978-981-99-8170-0_3

asynchronously determine whether an event is triggered between two neighboring agents. The guarantee of consensus and exclusion of Zeno behavior are similarly discussed.

Compared to those in the existing works, e.g., [16, 17, 28, 87, 102, 113, 119], both the edge-based and node-based adaptive event-triggered consensus protocols in this chapter are fully distributed and scalable, which do not rely on any global information of the network graph and are independent of the network's scale. Besides, these adaptive event-triggered protocols rely on the sampled local information of each agent and its neighbors, do not need continuous communications in either control laws updating or triggering conditions monitoring, and thereby can significantly reduce the communication frequency.

3.1 Problem Statement

Consider a group of identical agents with continuous-time general linear dynamics, described by

$$\dot{x}_i = Ax_i + Bu_i, \; i = 1, \cdots, N, \tag{3.1}$$

where $x_i \in \mathbf{R}^n$ and $u_i \in \mathbf{R}^p$ are, respectively, the state and control input of agent i, and A and B are constant matrices with compatible dimensions.

Assumption 3.1 The pair (A, B) is stabilizable.

Assumption 3.2 The graph $\mathcal{G}$ associated with the N agents is undirected and connected.

The objective of this chapter is to address the event-triggered consensus problem for the agents in (3.1), by ensuring that $\lim_{t\to\infty} \|x_i - x_j\| = 0, i, j = 1, \cdots, N$, and excluding the Zeno behavior, i.e., there does not exist an infinite number of events within a finite period of time [21, 102].

3.2 Linear Event-Triggered Consensus Protocol

In this section, we propose a linear fixed-gain event-based consensus protocol to achieve consensus for the agents in (3.1).

Let $\tilde{x}_i(t) \triangleq e^{A(t-t_k^i)} x_i(t_k^i), \forall t \in [t_k^i, t_{k+1}^i), i = 1, \cdots, N$, where t_k^i denotes the k-th event instant of agent i, with $t_0^i \triangleq 0$. For each agent, we define a measurement error $e_i(t)$ using the state error between the current time t and the latest event time t_k^i, i.e.,

$$e_i(t) \triangleq \tilde{x}_i(t) - x_i(t), \; i = 1, \cdots, N. \tag{3.2}$$

Based on the relative states of neighboring agents, we propose the following distributed control law using the event-triggered strategy:

$$u_i(t) = cK \sum_{j=1}^{N} a_{ij}(\tilde{x}_i(t) - \tilde{x}_j(t)),\ i = 1, \cdots, N, \tag{3.3}$$

where $c \in \mathbf{R}$ is a positive constant coupling gain, $K \in \mathbf{R}^{p\times n}$ is the feedback gain matrix, and a_{ij} denotes the (i, j)-th entry of the adjacency matrix associated with the graph $\mathcal{G}$.

A static triggering function for each agent i is given by

$$f_i(t) = 4d_i\|Ke_i\|^2 - \sum_{j=1}^{N} a_{ij}\|K(\tilde{x}_i - \tilde{x}_j)\|^2 - \mu e^{-\nu t},\ i = 1, \cdots, N, \tag{3.4}$$

where d_i is the in-degree of agent i, μ and ν are positive constants. As long as the triggering condition $f_i(t) \geq 0$ is fulfilled, an event is triggered for agent i. At this time instant, agent i updates its consensus control law (3.3) using its current state and broadcasts its current state to its neighbors. Meanwhile, the measurement error $e_i(t)$ is reset to zero. When agent i receives new information from its neighbors, the control law for agent i will also be updated immediately. If $f_i(t) < 0$ for $\forall i = 1, \cdots, N$, there is no communication occurring until the next event is triggered.

Remark 3.1 The event-triggered consensus protocol, composed of the consensus control laws (3.3) and the triggering functions (3.4), is based on the sampled states of each agent and those received from its neighbors, rather than the agent's real state. In the proposed protocol, each agent does not need to continuously monitor its neighbors' state and therefore the neighboring agents do not need continuous communication. In this way, this protocol can indeed reduce the communication frequency and save the limited resources of communication capacity.

Let $\xi = [\xi_1^T, \cdots, \xi_N^T]^T$ and $\tilde{\xi} = [\tilde{\xi}_1^T, \cdots, \tilde{\xi}_N^T]^T$, where $\xi_i = x_i - \frac{1}{N}\sum_{j=1}^{N} x_j$ and $\tilde{\xi}_i = \tilde{x}_i - \frac{1}{N}\sum_{j=1}^{N}\tilde{x}_j$, $i = 1, \cdots, N$. Then, we can get that $\xi = (M \otimes I_n)x$ and $\tilde{\xi} = (M \otimes I_n)\tilde{x}$, where $M = I_N - \frac{1}{N}\mathbf{1}\mathbf{1}^T$. It is clear that 0 is a simple eigenvalue of M with $\mathbf{1}$ as the corresponding right eigenvector and 1 is the other eigenvalue with multiplicity $N - 1$. Then, it follows that $\xi = 0$ if and only if $x_1 = \cdots = x_N$. Thus, we refer to ξ as the consensus error. Using (3.3) for (3.1), it follows that ξ satisfies the following closed-loop network dynamics:

$$\dot{\xi} = (I_N \otimes A)\xi + (c\mathcal{L} \otimes BK)\tilde{\xi}. \tag{3.5}$$

The following theorem shows that the proposed event-based protocol can indeed achieve consensus.

Theorem 3.1 *Suppose that Assumptions 3.1 and 3.2 hold. The consensus problem of the agents described by (3.1) is solved by the event-based consensus protocol composed of (3.3) and (3.4), with* $c \geq \frac{1}{\lambda_2(\mathcal{L})}$ *and* $K = -B^T P$, *where* $P > 0$ *is a solution to the following algebraic Riccati equation (ARE):*

$$PA + A^T P - PBB^T P + I = 0. \tag{3.6}$$

Proof Consider the following Lyapunov function candidate:

$$V_1 = \frac{1}{2}\xi^T (I_N \otimes P)\xi. \tag{3.7}$$

The time derivative of V_1 along the trajectory of (3.5) can be obtained as

$$\dot{V}_1 = \xi^T (I_N \otimes PA)\xi + \xi^T (c\mathcal{L} \otimes PBK)\tilde{\xi}. \tag{3.8}$$

Note that

$$\begin{aligned} \xi^T (c\mathcal{L} \otimes PBK)\tilde{\xi} = &-\frac{1}{2}\xi^T (c\mathcal{L} \otimes PBB^T P)\xi + \frac{1}{2}e^T (c\mathcal{L} \otimes PBB^T P)e \\ &- \frac{1}{2}\tilde{\xi}^T (c\mathcal{L} \otimes PBB^T P)\tilde{\xi}. \end{aligned} \tag{3.9}$$

Substituting (3.9) into (3.8) yields

$$\begin{aligned} \dot{V}_1 = &\frac{1}{2}\xi^T \left[I_N \otimes (PA + A^T P) - c\mathcal{L} \otimes PBB^T P\right]\xi \\ &+ \frac{1}{2}c\left[e^T (\mathcal{L} \otimes PBB^T P)e - \tilde{x}^T (\mathcal{L} \otimes PBB^T P)\tilde{x}\right]. \end{aligned} \tag{3.10}$$

Because $a_{ij} = a_{ji}$, it is not difficult to verify that

$$\begin{aligned} e^T (\mathcal{L} \otimes PBB^T P)e &= \sum_{i=1}^{N}\sum_{j=1}^{N} a_{ij} e_i^T PBB^T P(e_i - e_j) \\ &\leq \frac{3}{2}\sum_{i=1}^{N}\sum_{j=1}^{N} a_{ij} e_i^T PBB^T Pe_i + \frac{1}{2}\sum_{i=1}^{N}\sum_{j=1}^{N} a_{ij} e_j^T PBB^T Pe_j \\ &= 2\sum_{i=1}^{N} d_i \|Ke_i\|^2, \end{aligned} \tag{3.11}$$

and

$$\begin{aligned}\tilde{x}^T(\mathcal{L}\otimes PBB^TP)\tilde{x} &= \sum_{i=1}^{N}\sum_{j=1}^{N} a_{ij}\tilde{x}_i^T PBB^T P(\tilde{x}_i - \tilde{x}_j)\\ &= \frac{1}{2}\sum_{i=1}^{N}\sum_{j=1}^{N} a_{ij}\|K(\tilde{x}_i - \tilde{x}_j)\|^2.\end{aligned}\tag{3.12}$$

Substituting (3.11) and (3.12) into (3.10) yields

$$\begin{aligned}\dot{V}_1 &\le \frac{1}{2}\xi^T\left[I_N\otimes(PA+A^TP) - c\mathcal{L}\otimes PBB^TP\right]\xi\\ &\quad + \frac{c}{4}\sum_{i=1}^{N}\left[4d_i\|Ke_i\|^2 - \sum_{j=1}^{N} a_{ij}\|K(\tilde{x}_i-\tilde{x}_j)\|^2\right]\\ &\le \frac{1}{2}\xi^T\left[I_N\otimes(PA+A^TP) - c\mathcal{L}\otimes PBB^TP\right]\xi + \frac{Nc\mu}{4}e^{-\nu t},\end{aligned}\tag{3.13}$$

where we have used the triggering condition (3.4) to get the last inequality.

By the definition of ξ, it is easy to see that $(\mathbf{1}^T\otimes I)\xi = 0$. Because $\mathcal{G}$ is connected, it then follows from Lemma 2.2 that

$$\xi^T(\mathcal{L}\otimes PBB^TP)\xi \ge \lambda_2(\mathcal{L})\xi^T(I_N\otimes PBB^TP)\xi. \tag{3.14}$$

Because $c \ge \frac{1}{\lambda_2(\mathcal{L})}$, by substituting (3.6) and (3.14) into (3.13), we can obtain that

$$\begin{aligned}\dot{V}_1 &\le \frac{1}{2}\xi^T[I_N\otimes(PA+A^TP-PBB^TP)]\xi + \frac{c}{4}N\mu e^{-\nu t}\\ &\le -\frac{1}{2}\xi^T\xi + \frac{c}{4}N\mu e^{-\nu t}\\ &\le -\frac{1}{2}\xi^T\xi + \frac{c}{4}N\mu e^{-\nu t}.\end{aligned}\tag{3.15}$$

From (3.15), we can get that

$$0 \le V_1(t) \le \frac{c}{4}N\mu\int_0^t e^{-\nu\tau}d\tau.$$

Thus, V_1 is bounded. Invoking (3.7), ξ is bounded, which further implies that $\dot{\xi}$ is bounded according to (3.5). Besides, (3.15) enforces that

$$V_1(\infty) - V_1(0) \leq -\frac{1}{2}\int_0^\infty \xi^T(\tau)\xi(\tau)d\tau + \frac{cN\mu}{4\nu}. \tag{3.16}$$

Note that (3.16) can be rewritten as

$$\int_0^\infty \xi^T(\tau)\xi(\tau)d\tau \leq 2\left[V_1(\infty) - V_1(0) + \frac{cN\mu}{4\nu}\right].$$

Thus, using Lemma 2.4 gives that $\xi(t) \to 0$ as $t \to \infty$, i.e., consensus is achieved. □

Remark 3.2 Event-triggered consensus problems have been studied in previous related works, e.g., [21, 69, 87, 102, 119], where several event-based consensus protocols have been proposed. It should be pointed out that the nonzero eigenvalues of the Laplacian matrix associated with the communication graph, which are actually globe information of the network, are required to determine some parameters of the proposed event-based protocol (3.3) and (3.4) or those in the aforementioned works. Therefore, these event-based consensus protocols are not fully distributed. This motivates us to remove the limitation by presenting fully distributed and scalable event-triggered consensus schemes. Since the protocols are expected to be independent of the scale of network, the existing approach, e.g., in [87, 102], which estimates the lower bound of the nonzero eigenvalue of Laplacian matrix, is not applicable anymore. In the following sections, we will come up with novel perspectives to design fully distributed and scalable event-based protocols.

3.3 Fully Distributed Adaptive Event-Triggered Consensus Protocol: Type I

Inspired by the adaptive consensus protocols with continuous communications in [54, 55], we propose a distributed node-based event-triggered adaptive control law for each agent as follows:

$$\begin{aligned} u_i(t) &= K\sum_{j=1}^{N} c_{ij}(t)a_{ij}(\tilde{x}_i(t) - \tilde{x}_j(t)),\ i = 1, \cdots, N, \\ \dot{c}_{ij}(t) &= \kappa_{ij}a_{ij}[-\varrho_{ij}c_{ij}(t) + (\tilde{x}_i(t) - \tilde{x}_j(t))^T\Gamma(\tilde{x}_i(t) - \tilde{x}_j(t))], \end{aligned} \tag{3.17}$$

where $c_{ij}(t)$ denotes the time-varying coupling weight for the edge (i, j) with $c_{ij}(0) = c_{ji}(0)$, $\varrho_{ij} = \varrho_{ji}$ and $\kappa_{ij} = \kappa_{ji}$ are positive constants, and $K \in \mathbf{R}^{p\times n}$ and $\Gamma \in \mathbf{R}^{n\times n}$ are the feedback gain matrices.

The triggering function for each agent in this case is given by

$$f_i(t) = \sum_{j=1}^{N}(1+\delta c_{ij})a_{ij}e_i^T\Gamma e_i - \frac{1}{4}\sum_{j=1}^{N}a_{ij}(\tilde{x}_i - \tilde{x}_j)^T\Gamma(\tilde{x}_i - \tilde{x}_j) - \mu e^{-\nu t}, \tag{3.18}$$

where δ, μ, and ν are positive constants. The event triggering instant is defined as $t_{k+1}^i \triangleq \inf\{t > t_k^i | f_i(t) \geq 0\}$ with $t_0^i \triangleq 0$. The triggering function (3.18) works similarly as that in (3.4).

Remark 3.3 One distinct feature of the adaptive event-triggered protocol in this section is that it includes time-varying weights $c_{ij}(t)$ into both the control law (3.17) and the triggering function (3.18). As a consequence, the triggering function (3.18) here is non-quadratic in terms of the measurement error e_i, different from those in the previous works [87, 102, 119]. The triggering function (3.18) consists of a state-dependent term (i.e., the second term on the right-hand side) and a time-dependent term (i.e., the last exponential decay term). Similarly as in [74], we call (3.18) a hybrid or mixed triggering function. Combining the state-dependent and time-dependent terms in (3.18) is expected to be able to rule out the Zeno behavior and meanwhile decrease the event triggering number. Note that the event-based protocol, composed of (3.17) and (3.18), is model-based [28], and relies on the sampled state information of each agent and from its neighbors, rather than the agents' real states. Each agent does not need to continuously monitor its neighbors' states and therefore the neighboring agents do not need continuous communications.

Using (3.17) for (3.1), it follows that the consensus error ξ satisfies the following dynamics:

$$\begin{aligned} \dot{\xi}_i &= A\xi_i + BK\sum_{j=1}^{N} c_{ij}(t)a_{ij}(\tilde{x}_i - \tilde{x}_j), \\ \dot{c}_{ij}(t) &= \kappa_{ij}a_{ij}[-\varrho_{ij}c_{ij}(t) + (\tilde{x}_i - \tilde{x}_j)^T\Gamma(\tilde{x}_i - \tilde{x}_j)]. \end{aligned} \tag{3.19}$$

In what follows, we present an algorithm to construct the proposed event-triggered consensus protocol.

Algorithm 3.1 Assuming that Assumptions 3.1 and 3.2 hold, the event-triggered adaptive consensus protocol consisting of (3.17) and (3.18) can be designed according to the following three steps:

(1) Solve the ARE (3.6) to get a solution $P > 0$.
(2) Choose the feedback matrices $K = -B^TP$ and $\Gamma = PBB^TP$.
(3) Select κ_{ij}, ϱ_{ij}, δ, μ, and ν to be any positive constants.

We are now ready to present the main results of this section.

Theorem 3.2 *Suppose that Assumptions 3.1 and 3.2 hold. Both the consensus error* ξ *and the coupling gains* c_{ij} *in (3.19) are uniformly ultimately bounded under the event-triggered adaptive protocol constructed by Algorithm 3.1. If* ϱ_{ij} *in (3.17) are chosen such that* $\varrho_{ij}\kappa_{ij} < 1/\lambda_{\max}(P)$, *then* ξ *exponentially converges to a small adjustable bounded set as given in (3.34).*

Proof Consider the following Lyapunov function candidate:

$$V_2 = \frac{1}{2}\sum_{i=1}^{N} \xi_i^T P \xi_i + \sum_{i=1}^{N}\sum_{j=1, j\neq i}^{N} \frac{(c_{ij}-\alpha)^2}{8\kappa_{ij}}, \tag{3.20}$$

where α is a positive constant to be determined later. The time derivative of V_2 along the trajectory of (3.19) is given by

$$\begin{aligned}
\dot{V}_2 &= \sum_{i=1}^{N} \xi_i^T P \dot{\xi}_i + \sum_{i=1}^{N}\sum_{j=1, j\neq i}^{N} \frac{c_{ij}-\alpha}{4\kappa_{ij}} \dot{c}_{ij} \\
&= \sum_{i=1}^{N} \xi_i^T P A \xi_i + \sum_{i=1}^{N} \xi_i^T PBK \sum_{j=1}^{N} c_{ij} a_{ij} (\tilde{x}_i - \tilde{x}_j) \\
&\quad + \sum_{i=1}^{N}\sum_{j=1, j\neq i}^{N} \frac{c_{ij}-\alpha}{4\kappa_{ij}} \dot{c}_{ij}.
\end{aligned} \tag{3.21}$$

Since $a_{ij} = a_{ji}$ and $c_{ij}(t) = c_{ji}(t)$, it can be easily verified that

$$\sum_{i=1}^{N} \xi_i^T PBK \sum_{j=1}^{N} c_{ij} a_{ij} (\tilde{x}_i - \tilde{x}_j) = -\frac{1}{2}\sum_{i=1}^{N}\sum_{j=1}^{N} c_{ij} a_{ij} (\xi_i - \xi_j)^T \Gamma (\tilde{x}_i - \tilde{x}_j). \tag{3.22}$$

By substituting (3.22) into (3.21), we have

$$\begin{aligned}
\dot{V}_2 &= \sum_{i=1}^{N} \xi_i^T P A \xi_i - \frac{1}{2}\sum_{i=1}^{N}\sum_{j=1}^{N} c_{ij} a_{ij} (\tilde{x}_i - \tilde{x}_j)^T \Gamma (\tilde{x}_i - \tilde{x}_j) \\
&\quad + \frac{1}{2}\sum_{i=1}^{N}\sum_{j=1}^{N} c_{ij} a_{ij} (e_i - e_j)^T \Gamma (\tilde{x}_i - \tilde{x}_j) \\
&\quad + \sum_{i=1}^{N}\sum_{j=1, j\neq i}^{N} \frac{c_{ij}-\alpha}{4\kappa_{ij}} \dot{c}_{ij},
\end{aligned} \tag{3.23}$$

where we have used the facts that $\xi_i - \xi_j = x_i - x_j$ and $e_i = \tilde{x}_i - x_i$. In light of Lemma 2.5, it is not difficult to obtain that

$$\sum_{i=1}^{N}\sum_{j=1}^{N} c_{ij}a_{ij}(e_i - e_j)^T\Gamma(\tilde{x}_i - \tilde{x}_j) \leq \frac{1}{2}\sum_{i=1}^{N}\sum_{j=1}^{N} c_{ij}a_{ij}(e_i - e_j)^T\Gamma(e_i - e_j) + \frac{1}{2}\sum_{i=1}^{N}\sum_{j=1}^{N} c_{ij}a_{ij}(\tilde{x}_i - \tilde{x}_j)^T\Gamma(\tilde{x}_i - \tilde{x}_j), \tag{3.24}$$

and

$$\begin{aligned}\sum_{i=1}^{N}\sum_{j=1}^{N} c_{ij}a_{ij}(e_i - e_j)^T\Gamma(e_i - e_j) &\leq 2\sum_{i=1}^{N}\sum_{j=1}^{N} c_{ij}a_{ij}e_i^T\Gamma e_i \\ &\quad + 2\sum_{i=1}^{N}\sum_{j=1}^{N} c_{ij}a_{ij}e_j^T\Gamma e_j \\ &= 4\sum_{i=1}^{N}\sum_{j=1}^{N} c_{ij}a_{ij}e_i^T\Gamma e_i.\end{aligned} \tag{3.25}$$

Substituting (3.19), (3.24), and (3.25) into (3.23) yields

$$\begin{aligned}\dot{V}_2 &\leq \sum_{i=1}^{N}\xi_i^T PA\xi_i - \frac{\alpha}{4}\sum_{i=1}^{N}\sum_{j=1}^{N} a_{ij}(\tilde{x}_i - \tilde{x}_j)^T\Gamma(\tilde{x}_i - \tilde{x}_j) \\ &\quad + \sum_{i=1}^{N}\sum_{j=1}^{N} c_{ij}a_{ij}e_i^T\Gamma e_i - \sum_{i=1}^{N}\sum_{j=1}^{N}\frac{c_{ij} - \alpha}{4}\varrho_{ij}a_{ij}c_{ij} \\ &\leq \sum_{i=1}^{N}\xi_i^T PA\xi_i - \frac{\alpha}{4}\sum_{i=1}^{N}\sum_{j=1}^{N} a_{ij}(\tilde{x}_i - \tilde{x}_j)^T\Gamma(\tilde{x}_i - \tilde{x}_j) \\ &\quad + \sum_{i=1}^{N}\sum_{j=1}^{N} c_{ij}a_{ij}e_i^T\Gamma e_i + \varsigma - \sum_{i=1}^{N}\sum_{j=1}^{N}\frac{\varrho_{ij}a_{ij}}{8}(c_{ij} - \alpha)^2,\end{aligned} \tag{3.26}$$

where we have used Lemma 2.5 to get the last inequality and $\varsigma = \sum_{i=1}^{N}\sum_{j=1}^{N}\frac{\varrho_{ij}a_{ij}}{8}\alpha^2$.

Note that

$$
\begin{aligned}
\sum_{i=1}^{N}\sum_{j=1}^{N} a_{ij}(\tilde{x}_i-\tilde{x}_j)^T\Gamma(\tilde{x}_i-\tilde{x}_j) &= \sum_{i=1}^{N}\sum_{j=1}^{N} a_{ij}(\xi_i-\xi_j)^T\Gamma(\xi_i-\xi_j) \\
&\quad + \sum_{i=1}^{N}\sum_{j=1}^{N} a_{ij}(e_i-e_j)^T\Gamma(e_i-e_j) \\
&\quad + 2\sum_{i=1}^{N}\sum_{j=1}^{N} a_{ij}(x_i-x_j)^T\Gamma(e_i-e_j),
\end{aligned}
\tag{3.27}
$$

$$
\begin{aligned}
-\sum_{i=1}^{N}\sum_{j=1}^{N} a_{ij}(x_i-x_j)^T\Gamma(e_i-e_j) &\le \frac{1}{4}\sum_{i=1}^{N}\sum_{j=1}^{N} a_{ij}(x_i-x_j)^T\Gamma(x_i-x_j) \\
&\quad + \sum_{i=1}^{N}\sum_{j=1}^{N} a_{ij}(e_i-e_j)^T\Gamma(e_i-e_j),
\end{aligned}
\tag{3.28}
$$

and

$$
\sum_{i=1}^{N}\sum_{j=1}^{N} a_{ij}(e_i-e_j)^T\Gamma(e_i-e_j) \le 4\sum_{i=1}^{N}\sum_{j=1}^{N} a_{ij}e_i^T\Gamma e_i. \tag{3.29}
$$

Substituting (3.27), (3.28), and (3.29) into (3.26) gives

$$
\begin{aligned}
\dot{V}_2 \le{}& \frac{1}{2}\xi^T\left[I_N\otimes(PA+A^TP)-\frac{\alpha}{4}\mathcal{L}\otimes\Gamma\right]\xi \\
&+\frac{\alpha}{2}\sum_{i=1}^{N}\left[\sum_{j=1}^{N}(1+\frac{2}{\delta\alpha}\cdot\delta c_{ij})a_{ij}e_i^T\Gamma e_i-\frac{1}{4}\sum_{j=1}^{N}a_{ij}(\tilde{x}_i-\tilde{x}_j)^T\Gamma(\tilde{x}_i-\tilde{x}_j)\right] \\
&-\sum_{i=1}^{N}\sum_{j=1}^{N}\frac{\varrho_{ij}a_{ij}}{8}(c_{ij}-\alpha)^2+\varsigma.
\end{aligned}
\tag{3.30}
$$

By the definition of ξ, it is easy to see that $(\mathbf{1}^T\otimes I)\xi=0$. Because $\mathcal{G}$ is connected, it then follows from Lemma 2.2 that $\xi^T(\mathcal{L}\otimes\Gamma)\xi\ge\lambda_2(\mathcal{L})\xi^T(I_N\otimes\Gamma)\xi$, where $\lambda_2(\mathcal{L})$ is the smallest nonzero eigenvalue of $\mathcal{L}$. By noting the triggering functions

(3.18) and choosing α to be sufficiently large such that $\alpha \geq \max\{\frac{2}{\delta}, \frac{4}{\lambda_2(\mathcal{L})}\}$, it follows from (3.30) that

$$\begin{aligned}\dot{V}_2 \leq& \frac{1}{2}\xi^T \left[I_N \otimes (PA + A^T P) - \frac{\alpha}{4}\mathcal{L} \otimes PBB^T P \right] \xi \\ &- \sum_{i=1}^N \sum_{j=1}^N \frac{\varrho_{ij} a_{ij}}{8} (c_{ij} - \alpha)^2 + \varsigma + \frac{\alpha}{2} N \mu e^{-\nu t} \\ \leq& -\frac{1}{2}\xi^T \xi - \sum_{i=1}^N \sum_{j=1}^N \frac{\varrho_{ij} a_{ij}}{8} (c_{ij} - \alpha)^2 + \varsigma + \frac{\alpha}{2} N \mu e^{-\nu t}.\end{aligned} \tag{3.31}$$

Substituting (3.20) into (3.31) yields

$$\begin{aligned}\dot{V}_2 \leq& -\theta_1 V_2 + \frac{1}{2}\theta_1 \xi^T (I_N \otimes P)\xi - \frac{1}{2}\xi^T \xi \\ &+ \frac{1}{8}\sum_{i=1}^N \sum_{j=1}^N (\frac{\theta_1}{\kappa_{ij}} - \varrho_{ij}) a_{ij} (c_{ij} - \alpha)^2 + \varsigma + \frac{\alpha}{2} N \mu e^{-\nu t} \\ \leq& -\theta_1 V_2 + \varsigma + \frac{\alpha}{2} N \mu e^{-\nu t},\end{aligned}$$

where $\theta_1 = \min_{(i,j)\in\mathcal{E}}\{\varrho_{ij}\kappa_{ij}, \frac{1}{\lambda_{\max}(P)}\}$. According to Lemma 2.6, we have

$$V_2(t) \leq [V_2(0) - \frac{\varsigma}{\theta_1}]e^{-\theta_1 t} + \frac{\varsigma}{\theta_1} + \frac{\alpha}{2} N \psi(t),$$

where $\psi(t)$ is defined as

$$\psi(t) = \begin{cases} \mu t e^{-\theta_1 t} & \text{if } \theta_1 = \nu, \\ \frac{\mu}{\theta_1 - \nu}(e^{-\nu t} - e^{-\theta_1 t}) & \text{if } \theta_1 \neq \nu. \end{cases}$$

It is not difficult to verify that $\lim_{t\to+\infty} \psi(t) = 0$. Therefore, V_2 exponentially converges to a bounded set:

$$\mathcal{S}_1 \triangleq \left\{ \xi,\ c_{ij} \mid V_2 \leq \frac{\varsigma}{\theta_1} \right\}. \tag{3.32}$$

In light of the fact that $V_2 \geq \frac{\lambda_{\min}(P)}{2}\|\xi\|^2 + \sum_{i=1}^N \sum_{j=1}^N \frac{(c_{ij}-\alpha)^2}{8\kappa_{ij}}$, we conclude that ξ and c_{ij} are all uniformly ultimately bounded.

Since $\theta_2 \triangleq \min_{\forall(i,j)\in\mathcal{E}}\{\varrho_{ij}\kappa_{ij}\} < \frac{1}{\lambda_{\max}(P)}$, we can rewrite (3.31) as

$$
\begin{aligned}
\dot{V}_2 &\leq -\theta_2 V_2 + \frac{1}{2}\theta_2 \xi^T (I_N \otimes P)\xi - \frac{1}{2}\xi^T\xi + \varsigma + \frac{\alpha}{2}N\mu e^{-\nu t} \\
&\quad + \frac{1}{8}\sum_{i=1}^{N}\sum_{j=1}^{N}(\frac{\theta_2}{\kappa_{ij}} - \varrho_{ij})a_{ij}(c_{ij} - \alpha)^2 \\
&\leq -\theta_2 V_2 - \rho\xi^T\xi + \varsigma + \frac{\alpha}{2}N\mu e^{-\nu t},
\end{aligned}
\tag{3.33}
$$

where $\rho = \frac{1}{2}(1 - \theta_2\lambda_{\max}(P))$. Obviously, it follows from (3.33) that $\dot{V}_2 \leq -\theta_2 V_2 + \frac{\alpha}{2}N\mu e^{-\nu t}$ if $\xi^T\xi > \frac{\varsigma}{\rho}$. Therefore, we can obtain that the consensus error ξ exponentially converges to the following bounded set:

$$
\mathcal{S}_2 \triangleq \left\{\xi \mid \|\xi\|^2 \leq \frac{\varsigma}{\rho}\right\}. \tag{3.34}
$$

This completes the proof. □

Theorem 3.2 shows that the consensus error ξ under (3.17) and (3.18) converges to a residual set that can be arbitrary small by choosing proper constants ϱ_{ij}. The term $-\varrho_{ij}c_{ij}(t)$ in (3.17) is inspired by the σ-modification technique in the adaptive control literature [44, 53]. When $-\varrho_{ij}c_{ij}(t)$ is removed from (3.17), the consensus error ξ will asymptotically converge to zero, as stated in the following corollary.

Corollary 3.1 *Let $\varrho_{ij} = 0$, $\forall(i,j) \in \mathcal{E}$ in (3.17). Under the conditions as in Theorem 3.2, the consensus error ξ asymptotically converges to zero.*

One advantage of the event-based adaptive protocol (3.17) including the term $-\varrho_{ij}c_{ij}(t)$ lies in that it is robust in presence of external disturbances or uncertainties. For instance, the robustness of (3.17) with respect to perturbed agents $\dot{x}_i = Ax_i + Bu_i + w_i$, where w_i are bounded disturbances, can be shown by following similar steps in the proof of Theorem 3.2 with few modifications. The upper bound of the consensus error in this case will depend on both ϱ_{ij} and the upper bounds of w_i. The details are skipped for brevity.

Remark 3.4 It is well known that a necessary and sufficient condition for the existence of a $P > 0$ to the ARE (3.6) is that (A, B) is stabilizable. Therefore, a sufficient condition for the existence of the adaptive protocol (3.17) and (3.18) satisfying Algorithm 3.1 and Theorem 3.2 is also that (A, B) is stabilizable.

Remark 3.5 The final consensus value $\varpi(t)$ reached by the agents can be established. Using (3.17) for (3.1), we obtain that $\dot{x} = (I_N \otimes A)x + (\mathcal{L}^c \otimes BK)\tilde{x}$, where $\mathcal{L}^c$ is defined as $\mathcal{L}^c_{ii} = \sum_{j=1,j\neq i}^{N} c_{ij}a_{ij}$ and $\mathcal{L}^c_{ij} = -c_{ij}a_{ij}$, $i \neq j$, and $\tilde{x} = [\tilde{x}_1^T, \cdots, \tilde{x}_N^T]^T$. Noting that $\mathcal{L}^c$ is a weighted symmetric Laplacian matrix of $\mathcal{G}$, it is not difficult to verify that $(\mathbf{1}^T \otimes e^{-At})x$ is an invariant quantity.

Therefore, $(\mathbf{1}^T \otimes e^{-At})(\mathbf{1} \otimes \varpi(t)) = (\mathbf{1}^T \otimes I)x_0$, from which we can derive that $\varpi(t) = \frac{1}{N}\sum_{i=1}^{N} e^{At}x_i(0)$.

Remark 3.6 In Theorem 3.2 and Corollary 3.1, the communication graph $\mathcal{G}$ is assumed to be fixed throughout the whole process. Actually, the proposed adaptive consensus protocol is applicable to the case of arbitrary switching communication graphs with a positive dwelling time, which are connected at each contiguous time interval. In this case, communications only take place when the triggering conditions are violated or the topology switches. It is not challenging to prove this assertion by taking V_2 in (3.20) as a common Lyapunov function.

Remark 3.7 Contrary to the protocols in previous works [17, 28, 87, 102, 119], which require global information of the communication graph in terms of the nonzero eigenvalues of the corresponding Laplacian matrix, the proposed event-based adaptive protocol, consisting of the control law (3.17) and the triggering function (3.18), is said to be fully distributed and scalable, since it relies on none global information of the network graph and is independent of the network's scale.

The following theorem excludes the Zeno behavior.

Theorem 3.3 *Under the conditions in Theorem 3.2, the network (3.19) does not exhibit the Zeno behavior and the interval between two consecutive triggering instants for any agent is strictly positive, as illustrated in (3.38).*

Proof For agent i, consider the evolution of $e_i(t)$ for $t \in [t_k^i, t_{k+1}^i)$, $t_{k+1}^i < \infty$. It follows from (3.1), (3.2), and (3.17) that

$$\dot{e}_i = Ae_i - \sum_{j=1}^{N} c_{ij}a_{ij}BK(\tilde{x}_i - \tilde{x}_j).$$

The time derivative of $\|e_i(t)\|$ for $t \in [t_k^i, t_{k+1}^i)$ can be written as

$$\begin{aligned}\frac{d\|e_i(t)\|}{dt} &= \frac{e_i^T}{\|e_i\|}\dot{e}_i \leq \|\dot{e}_i\| \\ &\leq \|A\|\|e_i\| + \sum_{j=1}^{N} c_{ij}a_{ij}\|BK\|\|\tilde{x}_i - \tilde{x}_j\|.\end{aligned} \tag{3.35}$$

As shown in Theorem 3.2, both $c_{ij}(t)$ and ξ are bounded, the latter of which implies that $x_i(t) - x_j(t)$, $\forall(i,j) \in \mathcal{E}$, is also bounded. Without loss of generality, assume that $c_{ij} \leq \bar{c}$ for some positive constant $\bar{c}$. Note that the interval between two consecutive triggering events is bounded. Thus, $e^{A(t-t_k^i)}$ is bounded for any $t \in [t_k^i, t_{k+1}^i)$. Since $e^{-At}\sum_{j=1}^{N} x_j$ is an invariant quantity (see Remark 3.5), we can derive from $x_i = \xi_i + \frac{1}{N}\sum_{j=1}^{N} x_j$ that $x(t)$ is finite for any finite t. Therefore, we can get that for any $t \in [t_k^i, t_{k+1}^i)$, $\tilde{x}_i - \tilde{x}_j = e^{A(t-t_k^i)}x_i(t_k^i) - e^{A(t-t_{k'}^j)}x_j(t_{k'}^j)$ is also bounded, where $t_{k'}^j \in [t_k^i, t_{k+1}^i)$. Then, it follows from (3.35) that

$$\frac{d\|e_i(t)\|}{dt} \leq \|A\|\|e_i\| + \bar{c}\sigma_i,$$

where σ_i denotes the upper bound of $\sum_{j=1}^N a_{ij}\|BK\|\|\tilde{x}_i - \tilde{x}_j\|$ for t from t_k^i to t_{k+1}^i. Consider a non-negative function $\varphi : [0, \infty) \to \mathbf{R}_{\geq 0}$, satisfying

$$\dot{\varphi} = \|A\|\varphi + \bar{c}\sigma_i, \ \varphi(0) = \|e_i(t_k^i)\| = 0. \tag{3.36}$$

Then, we have $\|e_i(t)\| \leq \varphi(t - t_k^i)$, where $\varphi(t)$ is the analytical solution to (3.36) and is given by $\varphi(t) = \frac{\bar{c}\sigma_i}{\|A\|}(e^{\|A\|t} - 1)$.

It is not difficult to see that the triggering function (3.18) satisfies $f_i(t) \leq 0$, if we have the following condition:

$$\|e_i\|^2 \leq \frac{\mu e^{-\nu t}}{d_i\|K\|^2(1 + \delta\bar{c})}. \tag{3.37}$$

In light of (3.37), it is easy to see that the interval between two triggering instants t_k^i and t_{k+1}^i for agent i can be lower bounded by the time for $\varphi^2(t - t_k^i)$ evolving from 0 to the right-hand side of (3.37). Therefore, a lower bound τ_k^i of $t_{k+1}^i - t_k^i$ can be obtained by solving the following inequality:

$$\frac{\bar{c}^2\sigma_i^2}{\|A\|^2}(e^{\|A\|\tau_k^i} - 1)^2 \geq \frac{\mu e^{-\nu(t_k^i + \tau_k^i)}}{d_i\|K\|^2(1 + \delta\bar{c})}.$$

Then, we get that

$$\begin{aligned} t_{k+1}^i - t_k^i &\geq \tau_k^i \\ &\geq \frac{1}{\|A\|}\ln\left(1 + \frac{\|A\|}{\bar{c}\sigma_i\|K\|}\sqrt{\frac{\mu e^{-\nu(t_k^i + \tau_k^i)}}{d_i(1 + \delta\bar{c})}}\right). \end{aligned} \tag{3.38}$$

Note that τ_k^i always exists and is strictly positive for any finite time. Since the right-hand side of the second inequality in (3.38) approaches zero only when $t \to \infty$, we get that $t_k^i \to \infty$ with $k \to \infty$ and $t \to \infty$. Therefore, the Zeno behavior is excluded for the agents for any finite time. □

Remark 3.8 It is worth mentioning that the proof of Theorem 3.3 is partly inspired by Cheng et al. [17], Garcia et al. [28], and Sun et al. [89]. Note that the lower bound τ_k^i for the inter-event intervals in (3.38) is generally conservative, since it is derived by only using the exponential decay term in the triggering function (3.18) and ignoring the effect of the state-dependent term. Nevertheless, an advantage of the state-dependent term in (3.18) is that it can significantly reduce the number of event triggering, which can also be verified by numerical simulations. Besides, the lower bound τ_k^i in (3.38) depends on the specific time instants, not uniform with respect to t, and approaches zero when $t \to \infty$. If we replace $\mu e^{-\nu t}$ in (3.18) by a small positive constant μ, then the lower bound τ_k^i for the inter-event intervals satisfies $\tau_k^i \geq \frac{1}{\|A\|} \ln(1 + \frac{\|A\|}{\bar{c}\sigma_i \|K\|} \sqrt{\frac{\mu}{d_i(1+\delta\bar{c})}})$, which is always positive for any time. The cost is that in this case the asymptotical convergence of the consensus error cannot be guaranteed and only practical consensus can be expected.

Now, we illustrate the effectiveness of the theoretical results by conducting numerical simulations. For illustration, consider the linear multi-agent systems described by (3.1) with

$$A = \begin{bmatrix} 0 & 1 & 0 \\ 0 & 0 & 1 \\ 0 & 0 & 0 \end{bmatrix}, \quad B = \begin{bmatrix} 0 \\ 0 \\ 1 \end{bmatrix}.$$

All initial values of the agents are randomly chosen. The communication graph among the agents is depicted in Fig. 3.1.

Solving the ARE (3.6) gives a solution:

$$P = \begin{bmatrix} 2.4142 & 2.4142 & 1.0000 \\ 2.4142 & 4.8284 & 2.4142 \\ 1.0000 & 2.4142 & 2.4142 \end{bmatrix}.$$

Thus, the feedback matrices in the event-based control law (3.17) can be obtained as

$$K = -\begin{bmatrix} 1 & 2.4142 & 2.4142 \end{bmatrix}, \quad \Gamma = \begin{bmatrix} 1.0000 & 2.4142 & 2.4142 \\ 2.4142 & 5.8284 & 5.8284 \\ 2.4142 & 5.8284 & 5.8284 \end{bmatrix}.$$

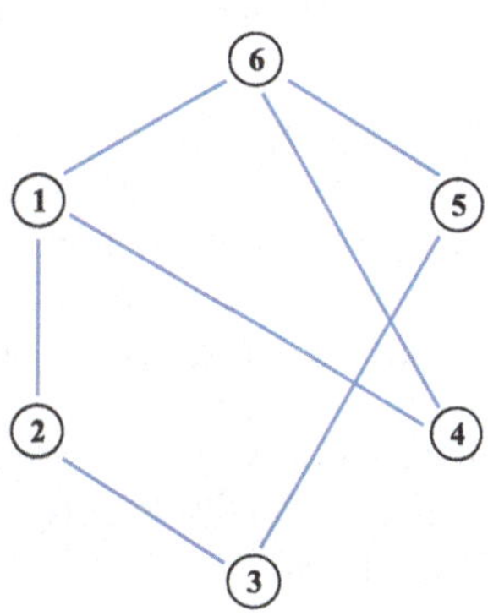

Fig. 3.1 The undirected communication graph among agents

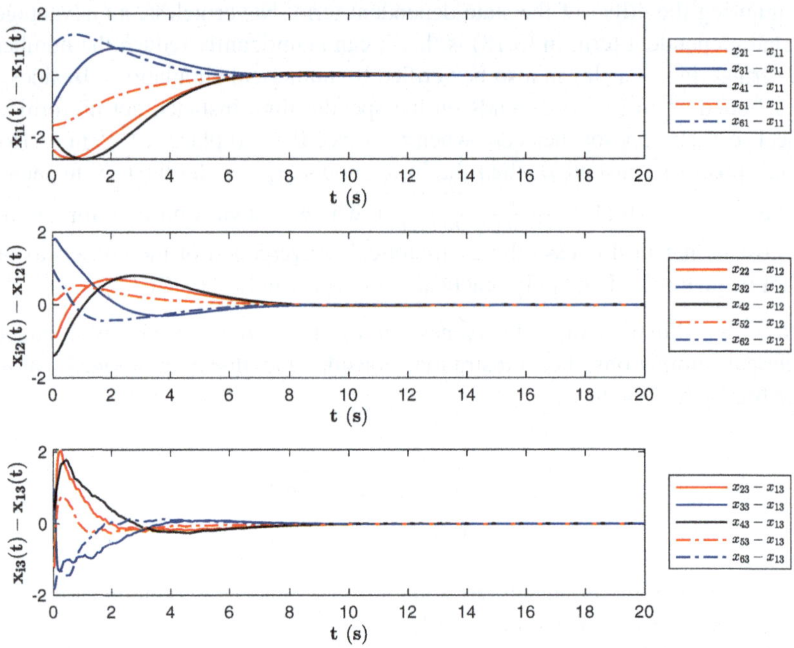

Fig. 3.2 The consensus errors $x_i - x_1$, $i = 2, \cdots, 6$

Other parameters of (3.17) and (3.18) are chosen as $\delta = 2$, $\mu = 3$, $\nu = 1$, $\varrho_{ij} = 0$, and $\kappa_{ij} = 0.2$, $\forall (i, j) \in \mathcal{E}$. The consensus errors $x_i - x_1$, $i = 2, \cdots, N$, are depicted in Fig. 3.2, from which we can observe that consensus is indeed achieved. The adaptive coupling weights $c_{ij}(t)$ in (3.17) are shown in Fig. 3.3, which implies that $c_{ij}(t)$ converge to finite steady-state values. The triggering instants of each agent are presented in Fig. 3.4, which shows that the Zeno behavior is excluded.

3.4 Fully Distributed Adaptive Event-Triggered Consensus Protocol: Type II

The adaptive event-triggered protocol in Sect. 3.3 is usually called node-based, in which when an event is triggered for an agent (i.e., a node in the graph) according to the triggering function in (3.18), this agent will simultaneously send its updated state to all its neighbors. In this section, we will present a novel fully distributed adaptive event-triggered protocol with edge-based event-triggered mechanisms, which can asynchronously determine whether an event is triggered for each edge in the network.

Let $\tilde{x}_{ij} = e^{A(t-t_k^{ij})} x_i(t_k^{ij})$, where t_k^{ij} denotes the k-th event instant of edge (i, j). Then, let $\tilde{\eta}_{ij} = \tilde{x}_{ij} - \tilde{x}_{ji}$. We propose the following edge-based adaptive event-

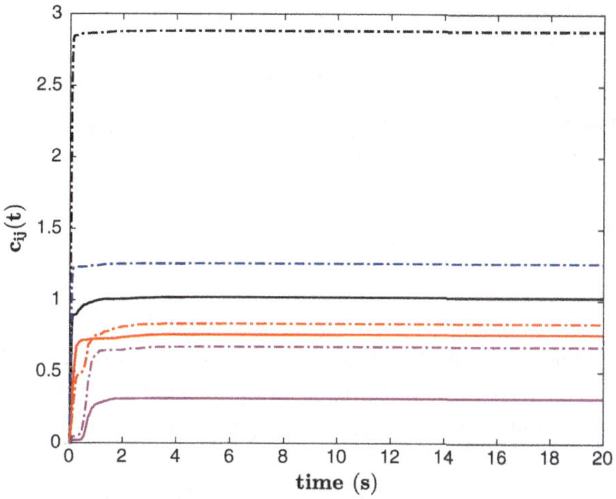

Fig. 3.3 The adaptive coupling weights $c_{ij}(t)$ in (3.17)

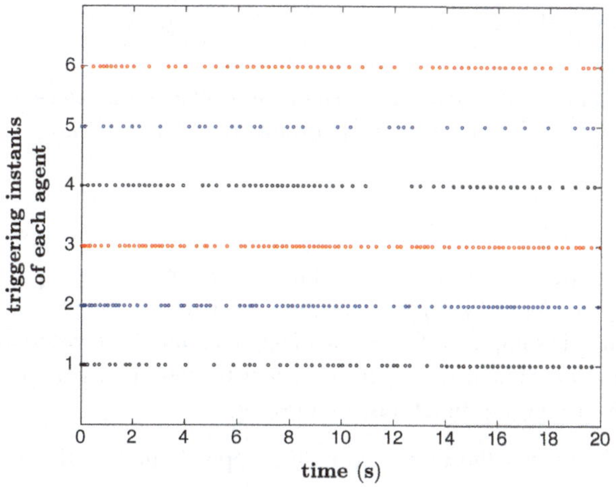

Fig. 3.4 Triggering instants of each agent

triggered consensus protocol:

$$u_i = K \sum_{j=1}^{N} c_{ij} a_{ij} \tilde{\eta}_{ij}, \tag{3.39a}$$

$$\dot{c}_{ij} = \kappa_{ij} a_{ij} \tilde{\eta}_{ij}^T \Gamma \tilde{\eta}_{ij}, \tag{3.39b}$$

where c_{ij} are the time-varying gains with $c_{ij}(0) = c_{ji}(0)$, $\kappa_{ij} = \kappa_{ji}$ are positive constants, and $K \in \mathbf{R}^{p\times n}$ and $\Gamma \in \mathbf{R}^{n\times n}$ are the feedback gain matrices.

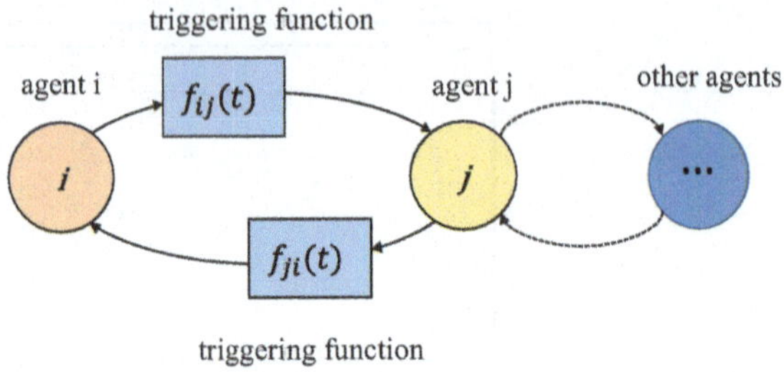

Fig. 3.5 The asynchronous edge-based event-triggered mechanism. The triggering function $f_{ij}(t)$ determines when to transmit the state information from agent i to agent j and update the state estimate $\tilde{x}_{ij}$ to $x_i(t)$

To determine the triggering instants t_k^{ij} for the edge (i, j), we need to define two triggering functions $f_{ij}(t)$ and $f_{ji}(t)$ for i and j, respectively, as shown in Fig. 3.5. The triggering functions $f_{ij}(t)$ and $f_{ji}(t)$ are given by

$$f_{ij}(t) = (1 + 2\delta c_{ij})e_{ij}^T \Gamma e_{ij} - \frac{1}{4}\tilde{\eta}_{ij}^T \Gamma \tilde{\eta}_{ij} - \mu_{ij} e^{-\nu_{ij} t}, \tag{3.40a}$$

$$f_{ji}(t) = (1 + 2\delta c_{ji})e_{ji}^T \Gamma e_{ji} - \frac{1}{4}\tilde{\eta}_{ji}^T \Gamma \tilde{\eta}_{ji} - \mu_{ji} e^{-\nu_{ji} t}, \tag{3.40b}$$

where δ, μ_{ij}, and ν_{ij} are positive constants and other parameters are the same as those in Sect. 3.3. Define the triggering instants for edge (i, j) as

$$t_{k+1}^{ij} \triangleq \{t_k^{ij} + \min\{\tau_k^{ij}, \tau\}\}, \; t_0^{ij} = t_0^{ji} \triangleq 0, \tag{3.41}$$

where τ is an upper bound of the event interval and $\tau_k^{ij} \triangleq \{t - t_k^{ij} > 0 \mid f_{ij}(t) \geq 0 \text{ or } f_{ji}(t) \geq 0\}$. Note that $f_{ij}(t)$ can be computed by agent i according to its information and the sampled information of agents i and j. Corresponding argument for the function $f_{ji}(t)$ is also true. In other words, continuous communications are not required for triggering functions' monitoring.

Remark 3.9 Note that the existence of the upper bound of the event interval τ provides convenience for the proof, specifically for the boundedness proof of the variable ξ. In real implementations, τ can be chosen as a very large value, and thus it will not bring conservatism for the event-triggered protocol. It should be noted that the idea of setting up an upper bound of the event interval is partly enlightened by Jain et al. [45].

Remark 3.10 Generally speaking, the triggering functions can be divided into two types, namely, node-based triggering and edge-based triggering. While both of them can be used to reduce the communication frequency, there exist some differences between them. Under the node-based event-triggered protocols in the previous section and also in other works such as [16, 98, 107], a triggering node, like i, has to transmit its current state information to its all neighbors. If there exists a neighbor of agent i, called j, whose state x_j is equal to or very near x_i, the communication between agents i and j at this instant is actually otiose. Different from the node-based event-triggered protocols, under edge-based ones, if an edge, like (i, j), is triggered, only agents i and j need to exchange their state information and update their control inputs. Intuitively, the edge-based event-triggered protocols proposed in this section are more propitious to reduce the updating frequency and save energy. On the other hand, the edge-based triggering functions are easier to compute compared with the node-based ones. In particular, for edge (i, j), the function $f_{ij}(t)$ depends on the relative information between agents i and j, rather than the sum of relative information between i and all its neighbors, which has to be computed in node-based ones.

Substituting (3.39) into (3.1) gives the dynamics of the consensus error ξ as follows:

$$\begin{aligned}\dot{\xi}_i &= A\xi_i + BK\sum_{j=1}^{N} c_{ij}a_{ij}\tilde{\eta}_{ij},\\ \dot{c}_{ij} &= \kappa_{ij}a_{ij}\tilde{\eta}_{ij}^T\Gamma\tilde{\eta}_{ij}.\end{aligned} \tag{3.42}$$

Theorem 3.4 *Suppose that Assumptions 3.1 and 3.2 hold. The consensus error ξ in (3.42) asymptotically converges to 0 under the proposed adaptive edge-based event-triggered protocol with $K = -B^TP$ and $\Gamma = PBB^TP$, where P satisfies (3.6).*

Proof Considering the Lyapunov function candidate V_2 defined in (3.20), we have

$$\dot{V}_2 = \sum_{i=1}^{N}\xi_i^TPA\xi_i + \sum_{i=1}^{N}\sum_{j=1}^{N}c_{ij}a_{ij}\xi_i^TPBK\tilde{\eta}_{ij} + \sum_{i=1}^{N}\sum_{j=1,j\neq i}^{N}\frac{c_{ij}-\alpha}{4\kappa_{ij}}\dot{c}_{ij}. \tag{3.43}$$

Noting that $c_{ij}(t) = c_{ji}(t)$, $(i, j) \in \mathcal{E}$, we can verify that

$$\begin{aligned}\sum_{i=1}^{N}\sum_{j=1}^{N}c_{ij}a_{ij}\xi_i^TPBK\tilde{\eta}_{ij} &= -\frac{1}{2}\sum_{i=1}^{N}\sum_{j=1}^{N}c_{ij}a_{ij}(\xi_i-\xi_j)^T\Gamma\tilde{\eta}_{ij}\\ &= -\frac{1}{2}\sum_{i=1}^{N}\sum_{j=1}^{N}c_{ij}a_{ij}\tilde{\eta}_{ij}^T\Gamma\tilde{\eta}_{ij}\\ &\quad+\frac{1}{2}\sum_{i=1}^{N}\sum_{j=1}^{N}c_{ij}a_{ij}(e_{ij}-e_{ji})^T\Gamma\tilde{\eta}_{ij}\\ &\leq -\frac{1}{4}\sum_{i=1}^{N}\sum_{j=1}^{N}c_{ij}a_{ij}\tilde{\eta}_{ij}^T\Gamma\tilde{\eta}_{ij} + \sum_{i=1}^{N}\sum_{j=0}^{N}c_{ij}a_{ij}e_{ij}^T\Gamma e_{ij},\end{aligned} \tag{3.44}$$

where we have used Lemma 2.5 to get the last inequality.

Using (3.39b), we have

$$\sum_{i=1}^{N}\sum_{j=1,j\neq i}^{N}\frac{c_{ij}-\alpha}{4\kappa_{ij}}\dot{c}_{ij}=\sum_{i=1}^{N}\sum_{j=1,j\neq i}^{N}a_{ij}\frac{c_{ij}-\alpha}{4}\tilde{\eta}_{ij}^{T}\Gamma\tilde{\eta}_{ij}. \tag{3.45}$$

It is not difficult to verify that

$$\begin{aligned}
-\frac{\alpha}{8}\sum_{i=1}^{N}\sum_{j=1,j\neq i}^{N}a_{ij}\tilde{\eta}_{ij}^{T}\Gamma\tilde{\eta}_{ij}&=-\frac{\alpha}{8}\sum_{i=1}^{N}\sum_{j=1}^{N}a_{ij}(\xi_i-\xi_j)^{T}\Gamma(\xi_i-\xi_j)\\
&\quad-\frac{\alpha}{4}\sum_{i=1}^{N}\sum_{j=1}^{N}a_{ij}(\xi_i-\xi_j)\Gamma(e_{ij}-e_{ji})\\
&\quad-\frac{\alpha}{8}\sum_{i=1}^{N}\sum_{j=1}^{N}a_{ij}(e_{ij}-e_{ji})^{T}\Gamma(e_{ij}-e_{ji})\\
&\leq-\frac{\alpha}{16}\sum_{i=1}^{N}\sum_{j=1}^{N}a_{ij}(\xi_i-\xi_j)^{T}\Gamma(\xi_i-\xi_j)\\
&\quad+\frac{\alpha}{8}\sum_{i=1}^{N}\sum_{j=1}^{N}a_{ij}(e_{ij}-e_{ji})^{T}\Gamma(e_{ij}-e_{ji})\\
&\leq-\frac{\alpha}{8}\sum_{i=1}^{N}\sum_{j=1}^{N}a_{ij}\xi_i^{T}\Gamma(\xi_i-\xi_j)+\frac{\alpha}{2}\sum_{i=1}^{N}\sum_{j=1}^{N}a_{ij}e_{ij}^{T}\Gamma e_{ij}.
\end{aligned}$$

Then, (3.45) can be simplified as

$$\begin{aligned}
\sum_{i=1}^{N}\sum_{j=1,j\neq i}^{N}\frac{c_{ij}-\alpha}{4\kappa_{ij}}\dot{c}_{ij}&\leq\frac{\alpha}{2}\sum_{i=1}^{N}\sum_{j=1}^{N}a_{ij}e_{ij}^{T}\Gamma e_{ij}-\frac{\alpha}{8}\sum_{i=1}^{N}\sum_{j=1}^{N}a_{ij}\xi_i^{T}\Gamma(\xi_i-\xi_j)\\
&\quad-\frac{\alpha}{8}\sum_{i=1}^{N}\sum_{j=1}^{N}a_{ij}\tilde{\eta}_{ij}^{T}\Gamma\tilde{\eta}_{ij}+\frac{1}{4}\sum_{i=1}^{N}\sum_{j=1}^{N}c_{ij}a_{ij}\tilde{\eta}_{ij}^{T}\Gamma\tilde{\eta}_{ij}.
\end{aligned} \tag{3.46}$$

Substituting (3.44) and (3.46) into (3.43) yields

$$\begin{aligned}
\dot{V}_2&\leq\xi^{T}(I_N\otimes PA)\xi-\frac{\alpha}{8}\xi^{T}(\mathcal{L}\otimes\Gamma)\xi\\
&\quad+\frac{\alpha}{2}\sum_{i=1}^{N}\sum_{j=1}^{N}a_{ij}\Big[(1+2\delta c_{ij})e_{ij}^{T}\Gamma e_{ij}-\frac{1}{4}\tilde{\eta}_{ij}^{T}\Gamma\tilde{\eta}_{ij}\Big].
\end{aligned} \tag{3.47}$$

Letting $\alpha = \max\{\frac{4}{\lambda_2(\mathcal{L})}, \frac{1}{\delta}\}$ and using the triggering functions (3.40) for (3.47), we have

$$\dot{V}_2 \leq -\frac{1}{2}\xi^T\xi + \varphi_2(t), \tag{3.48}$$

where $\varphi_2(t) = \frac{\alpha}{2}\sum_{i=1}^{N}\sum_{j=1}^{N} a_{ij}\mu_{ij}e^{-\nu_{ij}t}$.

On one hand, it follows from (3.48) that

$$0 \leq V_2(t) \leq V_2(0) + \varphi_3,$$

where $\varphi_3 = \frac{\alpha}{2}\sum_{i=1}^{N}\sum_{j=1}^{N} a_{ij}\frac{\mu_{ij}}{\nu_{ij}}$. In other words, V_2 is bounded, which further implies that ξ and c_{ij} are all bounded.

Next, we show the boundedness of $\dot{\xi}$. According to (3.42), we have

$$\dot{\xi}_i = A\xi_i + \sum_{j=1}^{N} c_{ij}a_{ij}BK(\tilde{x}_{ij} - \tilde{x}_{ji}).$$

Note that

$$\begin{aligned} \tilde{x}_{ij}(t) - \tilde{x}_{ji}(t) &= e^{A(t-t_k^{ij})}x_i(t_k^{ij}) - e^{A(t-t_k^{ij})}x_j(t_k^{ij}) \\ &= e^{A(t-t_k^{ij})}\left[x_i(t_k^{ij}) - x_j(t_k^{ij})\right]. \end{aligned} \tag{3.49}$$

Because ξ is bounded, $x_i(t_k^{ij}) - x_j(t_k^{ij})$ is also bounded. Noting that the event interval is upper bounded by τ according to the triggering rule (3.41), we have that $t - t_k^{ij}$ is bounded. Thus, $\tilde{x}_{ij}(t) - \tilde{x}_{ji}(t)$ is bounded and so is $\dot{\xi}$.

On the other hand, according to (3.48), we have

$$\begin{aligned} \int_0^{+\infty} \xi^T(t)\xi(t)dt &\leq \int_0^{+\infty} 2(-\dot{V}_2(t) + \varphi_2(t))dt \\ &= 2\left(V_2(0) - V_2(+\infty)\right) + 2\varphi_3. \end{aligned}$$

Thus, we can use Lemma 2.4 to conclude that $\xi(t) \to 0$ as $t \to +\infty$. This completes the proof. □

Theorem 3.5 *The Zeno behavior in the closed-loop network (3.42) is excluded.*

Proof A contradiction argument is provided to exclude the Zeno behavior. Without loss of generality, suppose that the edge (i, j) exhibits the Zeno behavior. According to the definition of Zeno behavior, $\exists T < +\infty \Rightarrow \lim_{k\to\infty} t_k^{ij} = T$. Then, for any $\varepsilon_0 > 0$, $\exists M_0 \in \mathbf{Z}_{\geq 0} \Rightarrow T - \varepsilon_0 < t_m^{ij} \leq T$, for any $m \geq M_0$.

Note that $\tilde{\eta}_{ij} = \tilde{x}_{ij}(t) - \tilde{x}_{ji}(t) = e^{A(t-t_k^{ij})}\left[x_i(t_k^{ij}) - x_j(t_k^{ij})\right]$. Because ξ is bounded, $x_i(t_k^{ij}) - x_j(t_k^{ij})$ is also bounded. Then, it follows from (3.49) that $\tilde{\eta}_{ij}$ is bounded in finite time. Compute the right-hand Dini derivative of e_{ij} as follows:

$$D^+ e_{ij} = Ae_{ij} - BK \sum_{j=0}^{N} c_{ij} a_{ij} \tilde{\eta}_{ij}.$$

Since we have shown that c_{ij} are all bounded, there exists a positive constant $\bar{c}$ satisfying $c_{ij} \leq \bar{c}$. It follows that

$$D^+ \|e_{ij}\| \leq \|A\| \|e_{ij}\| + \sigma_{ij},$$

where σ_{ij} represents the upper bound of $\left\| BK \sum_{j=1}^{N} c_{ij} a_{ij} \tilde{\eta}_{ij} \right\|$ for $t \in [t_k^{ij}, t_{k+1}^{ij})$.

Define a non-negative function $\phi : [0, \infty) \to \mathbf{R}_{\geq 0}$, satisfying

$$\dot{\phi} = \|A\| \phi + \sigma_{ij}, \; \phi(0) = \|e_{ij}(t_k^{ij})\| = 0. \tag{3.50}$$

Therefore, $\|e_{ij}(t)\| \leq \phi(t - t_k^{ij})$, where $\phi(t) = \frac{\sigma_{ij}}{\|A\|}\left(e^{\|A\|t} - 1\right)$ satisfies (3.50).

Note that the edge (i, j) is not triggered, i.e., the triggering functions (3.40) are no more than 0, if the following two inequalities hold:

$$\|e_{ij}\| \leq \sqrt{\frac{\mu_{ij}}{\|PBB^T P\|}} e^{-\frac{\nu_{ij}}{2}T}, \tag{3.51a}$$

$$\|e_{ji}\| \leq \sqrt{\frac{\mu_{ji}}{\|PBB^T P\|}} e^{-\frac{\nu_{ji}}{2}T}. \tag{3.51b}$$

Let

$$\varepsilon_{ij} = \frac{1}{2\|A\|} \ln \left(1 + \frac{\|A\|}{\sigma_{ij}} \sqrt{\frac{\mu_{ij}}{\|PBB^T P\|}} e^{-\frac{\nu_{ij}}{2}T}\right)$$

and

$$\varepsilon_{ji} = \frac{1}{2\|A\|} \ln \left(1 + \frac{\|A\|}{\sigma_{ji}} \sqrt{\frac{\mu_{ji}}{\|PBB^T P\|}} e^{-\frac{\nu_{ji}}{2}T}\right).$$

Denote $\varepsilon_0 \triangleq \min\{\varepsilon_{ij}, \varepsilon_{ji}\}$. Then, according to (3.51), the time interval between t_k^{ij} and t_{k+1}^{ij} satisfies $t_{k+1}^{ij} - t_k^{ij} \geq 2\varepsilon_0$, which further implies that $t_{k+1}^{ij} \geq t_k^{ij} + 2\varepsilon_0 > T + \varepsilon_0$. This contradicts the assumption that $T - \varepsilon_0 < t_{k+1}^{ij} \leq T$. Thus, there does not exist the Zeno behavior. □

Remark 3.11 It can be seen from Theorem 3.5 that the feasibility of the adaptive edge-based event-triggered protocols is guaranteed by providing a contradiction argument. In this section, we use another method to exclude the Zeno behavior, different from Sect. 3.3. Note that such two kinds of methods are often used to exclude the undesirable Zeno behavior in the literature, when studying the event-triggered consensus control problem of multi-agent systems.

3.5 Notes

The results of this chapter are mainly based on [12, 14], which, to the best of our knowledge, are the first to systematically study the fully distributed event-triggered consensus control problem. Both node-based and edge-based adaptive event-triggered protocols have been proposed to solve such a problem. These event-triggered protocols use only the local sampled information of each agent and its neighbors, and thereby can be designed and utilized in a fully distributed fashion. It should be pointed out that there are some other works considering the adaptive event-triggered control problem of multi-agent systems, e.g., [59, 81]. However, in those works, each agent needs to transmit the input information to its neighbors at event instants, which consumes extra computation cost and relies on the accuracy of system dynamics. Such limitations can be circumvented in this chapter.

Chapter 4
Fully Distributed Event-Triggered Consensus with Discrete Communication and Control

In Chap. 3, two types of fully distributed adaptive event-triggered consensus protocols have been presented, which do not rely on any global information of the network or continuous communications among neighboring agents. However, those protocols still suffer from two disadvantages. First, they are only applicable to undirected graphs or leader-follower graphs with undirected subgraphs among followers. How to extend them to the general directed graph case seems quite challenging or even impossible due to the special structures of the triggering functions or the control laws. Second, those protocols need continuous control updating, which is evidently uneconomic and undesirable, since this would give rise to the accelerated fatigue of actuators and the reduction of usable lives.

In this chapter, we extend to study the problem of fully distributed event-triggered consensus with both discrete communication and control (FDECDCC). Specifically, fully distributed event-triggered protocols hereby should simultaneously possess three features, namely, discrete communications among neighboring agents, discrete control updating of each agent, and independence of any global information of the network. The main challenge lies in the intensively coupled nonlinearity caused by the co-existence of the adaptive nonlinear control as well as the asynchronous and aperiodic discrete-time control updating and communication signals. General directed graphs also significantly increase the difficulty of this problem. To the best of our knowledge, until now many scientists have made contributions to only one or two requirements of this problem; see [21, 27, 36, 43, 69, 74, 87, 108] and the references therein. Even though some works including [29, 70] proposed several periodic event-triggered protocols, which either depend on global information or are only applicable to some simple cases such as single integrators or undirected graphs. In this chapter, we intend to systematically analyze and thoroughly work out this challenging FDECDCC problem.

In Sect. 4.1, we first formulate the FDECDCC problem. As explained earlier, those adaptive event-triggered protocols proposed in Chap. 3 cannot be used here and novel adaptive event-triggered protocols are expected. In Sect. 4.2, we first

Z. Li et al., *Distributed Event-triggered Control*,
https://doi.org/10.1007/978-981-99-8170-0_4

consider the simple case of undirected graphs and propose a novel event triggering functions, whose distinct feature, compared with the existing static triggering functions as in [12, 14, 59, 81] and in Chap. 3, is that they are dynamic and independent of any relative information among neighboring agents. Such dynamic triggering functions have the advantage that the local estimation error is decoupled from the network topology, providing opportunities to thoroughly solve the FDECDCC problem. Based on the proposed event triggering functions, a new framework of fully distributed adaptive event-triggered protocols is then presented, where both the control and communication are updated only at discrete event instants and are independent of any global information of the network. With the established framework, novel theoretical analysis method is established by introducing virtual adaptive gains to generate appropriate Lyapunov function candidates.

In Sect. 4.3, we further study the FDECDCC problem for the case with strongly connected directed graphs. Different from the undirected graph case in Sect. 4.2, the Laplacian matrix associated with the directed graph is asymmetric, which greatly increases the difficulty of protocol design and theoretical analysis. In order to deal with the unpleasant asymmetry caused by directed graphs, we present novel adaptive event-triggered protocols with time-varying coupling weights plus monotonically increasing functions. Consensus is achieved and Zeno behavior is excluded for the cases in Sects. 4.2 and 4.3.

We then extend the results of Sect. 4.3 to the case of graphs that contains only a directed spanning tree in Sect. 4.4. Two triggering functions are designed for the leader and the followers, respectively, to determine when to transmit information from the leader to informed followers, and when to carry out communications among neighboring followers. In Sect. 4.5, we consider the output feedback case. Both the control law and the triggering function in this case are designed based on the state estimate provided by the local Luenberger observer of each agent. In Sect. 4.6, we further propose fully distributed adaptive event-triggered protocols using the relative state measurement information. We also make some comparisons between these two different ideas of designing adaptive event-triggered protocols.

4.1 Problem Statement

Consider a group of N agents with dynamics described by

$$\begin{aligned} \dot{x}_i &= Ax_i + Bu_i, \\ y_i &= Cx_i, \quad i = 1, \cdots, N, \end{aligned} \tag{4.1}$$

where $x_i \in \mathbf{R}^n$, $y_i \in \mathbf{R}^m$, and $u_i \in \mathbf{R}^p$ represent the state, the measured output, and the control input of the i-th agent, respectively, and A, B, C are constant known matrices with compatible dimensions. Furthermore, the triple (A, B, C) is stabilizable and detectable. The communication topology among the N agents is represented by a directed graph $\mathcal{G}$.

In this chapter, we aim at studying the problem of FDECDCC. Three main constrains are involved in this problem, namely, limited updates of controllers, limited bandwidths of communications, and unavailable global information associated with the whole network. The above three constraints are very common in numerous practical occasions, while the co-presence of them renders the problem quite challenging. We shall design proper distributed protocols to realize consensus in the sense that $\lim_{t\to\infty}\|x_i(t)-x_j(t)\|=0, \forall i,j=1,\cdots,N$ and ensure that there does not exist the Zeno behavior [102].

Let $\xi=[\xi_1^T,\cdots,\xi_N^T]^T$, where $\xi_i=\sum_{j=1}^N a_{ij}(x_i-x_j)$. Then, we have $\xi=(\mathcal{L}\otimes I_n)x$, where $x=[x_1^T,\cdots,x_N^T]^T$. We refer to ξ as the consensus error. It is clear that consensus is achieved if and only if ξ asymptotically converges to zero.

4.2 Fully Distributed Event-Based Control for Undirected Graphs

In this section, we consider the FDECDCC problem under undirected graphs. Due to the constraint of discrete communication and discrete control updating, we cannot directly use the consensus error ξ_i to design the control input u_i as in the existing works like [53, 64]. Instead, only the state information at event-triggering instant can be utilized to generate the control commands. For convenience, denote $\hat{x}_j(t)=x_j(t_k^j), \forall t\in[t_k^j,t_{k+1}^j)$ as the state estimation of agent j, where t_k^j denotes the k-th event-triggering instant of agent j. Then, the main task is to co-design controllers and triggering functions based on $\hat{x}_j(t)$.

Defining $\hat{\xi}_i(t)=\sum_{j=1}^N a_{ij}(\hat{x}_i(t)-\hat{x}_j(t))$, we first design the following distributed node-based adaptive event-triggered controller for each agent:

$$\begin{aligned} u_i(t) &= \hat{d}_i(t)K\hat{\xi}_i(t), \\ \dot{d}_i(t) &= \hat{\xi}_i^T(t)\Gamma\hat{\xi}_i(t), \end{aligned} \tag{4.2}$$

where $\hat{d}_i(t)=d_i(t_k^i), \forall t\in[t_k^i,t_{k+1}^i)$, $d_i(t)$ is the adaptive coupling gain for agent i with initial value $d_i(0)\geq 1$, K and Γ are the feedback gain matrices to be designed.

The key now is to determine when to carry out communications and update controllers. Define $\tilde{x}_i(t)=\hat{x}_i(t)-x_i(t)$ and $\tilde{d}_i(t)=\hat{d}_i(t)-d_i(t)$ as the estimation errors. We design a triggering function for each agent as follows:

$$t_{k+1}^i=\inf\{t>t_k^i \mid \tilde{x}_i^T(t)\Gamma\tilde{x}_i(t)\geq\gamma_i\varepsilon_i(t) \ \vee\ |\tilde{d}_i|\geq\theta_1 e^{-\theta_2(t-t_k^i)}\}, \tag{4.3}$$

where $t_0^i = 0$, $\gamma_i > 0$, θ_1 and θ_2 are positive constants, and $\varepsilon_i(t)$ is an internal variable with dynamics:

$$\dot{\varepsilon}_i(t) = -k_i\varepsilon_i(t) - \sigma_i\tilde{x}_i^T(t)\Gamma\tilde{x}_i(t), \tag{4.4}$$

with $\varepsilon_i(0) > 0$, $k_i > 0$, and $\sigma_i > 0$. Only at event instant t_k^i, agent i sends its current state information to its out-neighbors. According to (4.2), agent i updates its control input u_i only when agent i or its neighboring agents are triggered.

Remark 4.1 It is not difficult to see from (4.2) that the control input u_i is designed completely based on the information at event instants and only updated at some discrete instants. To determine these event instants, we propose the novel dynamic rule composed of (4.3) and (4.4), from which, intuitively, the error $\tilde{x}_i$ and $\tilde{d}_i$ will not exceed the designed thresholds. Particularly, the so-called state-based threshold $-\sigma_i\tilde{x}_i^T(t)\Gamma\tilde{x}_i(t)$ only relies on agent i and is decoupled from the network topology. With such a decoupling feature, we can avoid both continuous communication and control updating.

Letting $\tilde{\xi}_i = \sum_{j=1}^N a_{ij}(\tilde{x}_i - \tilde{x}_j)$, we have $\tilde{\xi}_i = \hat{\xi}_i - \xi_i$. Denote $\bar{d}_i$ as the virtual adaptive gain, whose dynamics are given by

$$\dot{\bar{d}}_i = \xi_i^T\Gamma\xi_i,$$

with initial value $\bar{d}_i(0) \geq 1$. Further let $\check{d}_i = d_i - \bar{d}_i$. Then, we can derive the closed-loop dynamics of x_i as

$$\dot{x}_i = Ax_i + \bar{d}_iBK\xi_i + d_iBK\tilde{\xi}_i + \check{d}_iBK\xi_i + \tilde{d}_iBK\hat{\xi}_i.$$

The compact form of the dynamics of x can be written as

$$\dot{x} = (I_N \otimes A)x + (\bar{D} \otimes BK)\xi + (D \otimes BK)\tilde{\xi} + (\check{D} \otimes BK)\xi + (\tilde{D} \otimes BK)\hat{\xi}, \tag{4.5}$$

where $\xi = [\xi_1^T, \cdots, \xi_N^T]^T$, $\tilde{\xi} = [\tilde{\xi}_1^T, \cdots, \tilde{\xi}_N^T]^T$, $\hat{\xi} = [\hat{\xi}_1^T, \cdots, \hat{\xi}_N^T]^T$, $\bar{D} = \text{diag}\{\bar{d}_1, \cdots, \bar{d}_N\}$, $D = \text{diag}\{d_1, \cdots, d_N\}$, $\tilde{D} = \text{diag}\{\tilde{d}_1, \cdots, \tilde{d}_N\}$, and $\check{D} = \text{diag}\{\check{d}_1, \cdots, \check{d}_N\}$.

The following result shows the convergence of the consensus error ξ.

Theorem 4.1 *For the multi-agent system (4.1) under an undirected and connected graph, consensus can be achieved under the distributed adaptive event-triggered protocol (4.2) and (4.3) by choosing $K = -B^TQ$ and $\Gamma = QBB^TQ$ with $Q > 0$ whose inverse $P = Q^{-1}$ is the solution to the following linear matrix inequality (LMI):*

$$AP + PA^T - BB^T < 0. \tag{4.6}$$

Proof Consider the following Lyapunov function candidate:

$$V_1 = x^T(\mathcal{L} \otimes Q)x + \sum_{i=1}^{N} \left[\frac{1}{2}(\bar{d}_i - \alpha)^2 + \beta \bar{d}_i V_0 \right], \tag{4.7}$$

where α and β are positive constants to be determined, and

$$V_0 = \sum_{j=1}^{N} \frac{\varepsilon_j}{k_j}. \tag{4.8}$$

By (4.4), we have $-(k_i + \sigma_i \gamma_i)\varepsilon_i \leq \dot{\varepsilon}_i \leq -k_i \varepsilon_i$, and thus $e^{-(k_i+\sigma_i\gamma_i)t}\varepsilon_i(0) \leq \varepsilon_i(t) \leq e^{-k_i t}\varepsilon_i(0)$. The time derivative of V_1 along the trajectory of (4.5) is given by

$$\begin{aligned} \dot{V}_1 = {} & x^T[\mathcal{L} \otimes (QA + A^T Q)]x - 2\xi^T(\bar{D} \otimes \Gamma)\xi \\ & - 2\xi^T(D \otimes \Gamma)\tilde{\xi} - 2\xi^T(\check{D} \otimes \Gamma)\xi - 2\xi^T(\tilde{D} \otimes \Gamma)\tilde{\xi} \\ & + \sum_{i=1}^{N}(\bar{d}_i - \alpha + \beta V_0)\xi_i^T \Gamma \xi_i - \sum_{i=1}^{N} \beta \bar{d}_i \sum_{j=1}^{N} \left(\varepsilon_j + \frac{\sigma_j}{k_j} \tilde{x}_j^T \Gamma \tilde{x}_j \right). \end{aligned} \tag{4.9}$$

Specifically, the key is to analyze the three terms of (4.9), i.e., $-2\xi^T(D \otimes \Gamma)\tilde{\xi}$, $-2\xi^T(\check{D} \otimes \Gamma)\xi$, and $-2\xi^T(\tilde{D} \otimes \Gamma)\tilde{\xi}$.

(1) We first deal with the term $-2\xi^T(D \otimes \Gamma)\tilde{\xi}$. By Lemma 2.5, we have

$$\begin{aligned} -2\xi^T(D \otimes \Gamma)\tilde{\xi} & \leq \frac{1}{2}\xi^T(D \otimes \Gamma)\xi + 2\tilde{\xi}^T(D \otimes \Gamma)\tilde{\xi} \\ & = \frac{1}{2}\xi^T(\bar{D} \otimes \Gamma)\xi + \frac{1}{2}\xi^T(\check{D} \otimes \Gamma)\xi \\ & \quad + 2\tilde{\xi}^T(\bar{D} \otimes \Gamma)\tilde{\xi} + 2\tilde{\xi}^T(\check{D} \otimes \Gamma)\tilde{\xi}. \end{aligned}$$

Note that

$$\begin{aligned} \check{d}_i(t) & = \int_0^t (\hat{\xi}_i^T(\tau)\Gamma\hat{\xi}_i(\tau) - \xi_i^T(\tau)\Gamma\xi_i(\tau))d\tau + \check{d}_i(0) \\ & = \int_0^t (\tilde{\xi}_i^T(\tau)\Gamma\tilde{\xi}_i(\tau) + 2\xi_i^T(\tau)\Gamma\tilde{\xi}_i(\tau))d\tau + \check{d}_i(0) \\ & \leq \int_0^t \xi_i^T(\tau)\Gamma\xi_i(\tau)d\tau + 2\int_0^t \tilde{\xi}_i^T(\tau)\Gamma\tilde{\xi}_i(\tau)d\tau + \check{d}_i(0) \\ & = \bar{d}_i(t) + 2\int_0^t \tilde{\xi}_i^T(\tau)\Gamma\tilde{\xi}_i(\tau)d\tau + \check{d}_i(0) - \bar{d}_i(0), \end{aligned} \tag{4.10}$$

and

$$
\begin{aligned}
\tilde{\xi}_i^T \Gamma \tilde{\xi}_i &= \sum_{j=1}^N a_{ij}(\tilde{x}_i - \tilde{x}_j)^T \Gamma \sum_{j=1}^N a_{ij}(\tilde{x}_i - \tilde{x}_j) \\
&\le 2l_{ii}^2 \tilde{x}_i^T \Gamma \tilde{x}_i + 2\sum_{j=1}^N a_{ij}\tilde{x}_j^T \Gamma \sum_{j=1}^N a_{ij}\tilde{x}_j \qquad (4.11) \\
&\le 2l_{ii}^2 \tilde{x}_i^T \Gamma \tilde{x}_i + 2l_{ii}\sum_{j=1}^N a_{ij}\tilde{x}_j^T \Gamma \tilde{x}_j \le 2l_{ii}^2 \sum_{j=1}^N \tilde{x}_j^T \Gamma \tilde{x}_j.
\end{aligned}
$$

Combining (4.10) and (4.11), we can get that

$$
\begin{aligned}
\check{d}_i(t) &\le \bar{d}_i(t) - \bar{d}_i(0) + \check{d}_i(0) + 4l_{ii}^2 \sum_{j=1}^N \int_0^t \tilde{x}_j^T(\tau)\Gamma \tilde{x}_j(\tau)d\tau \\
&\le \bar{d}_i(t) - \bar{d}_i(0) + \check{d}_i(0) + 4l_{ii}^2 \sum_{j=1}^N \int_0^\infty \gamma_j \varepsilon_j(\tau)d\tau \qquad (4.12) \\
&\le \bar{d}_i(t) - \bar{d}_i(0) + \check{d}_i(0) + 4l_{ii}^2 \sum_{j=1}^N \frac{\gamma_j \varepsilon_j(0)}{k_j}.
\end{aligned}
$$

Then, we have

$$
2\tilde{\xi}^T(\check{D} \otimes \Gamma)\tilde{\xi} = \sum_{i=1}^N 2\check{d}_i \tilde{\xi}_i^T \Gamma \tilde{\xi}_i \le \sum_{i=1}^N \left(2\bar{d}_i + \beta_1\right) \tilde{\xi}_i^T \Gamma \tilde{\xi}_i,
$$

where $\beta_1 \ge \max\{-2\bar{d}_i(0) + 2\check{d}_i(0) + 8l_{ii}^2 \sum_{j=1}^N \frac{\gamma_j \varepsilon_j(0)}{k_j}\}$ is a positive constant. Therefore, we can derive that

$$
\begin{aligned}
-2\xi^T(D \otimes \Gamma)\tilde{\xi} &\le \frac{1}{2}\xi^T(\bar{D} \otimes \Gamma)\xi + \frac{1}{2}\xi^T(\check{D} \otimes \Gamma)\xi + \sum_{i=1}^N \left(4\bar{d}_i + \beta_1\right) \tilde{\xi}_i^T \Gamma \tilde{\xi}_i \\
&\le \frac{1}{2}\xi^T(\bar{D} \otimes \Gamma)\xi + \frac{1}{2}\xi^T(\check{D} \otimes \Gamma)\xi + \sum_{i=1}^N \left(8\bar{d}_i + 2\beta_1\right) l_{ii}^2 \sum_{j=1}^N \tilde{x}_j^T \Gamma \tilde{x}_j.
\end{aligned} \qquad (4.13)
$$

(2) We then deal with the term $-\frac{3}{2}\xi^T(\check{D}\otimes\Gamma)\xi$. Obviously, we have

$$\begin{aligned}-\check{d}_i(t) &= -\check{d}_i(0) + \int_0^t (\xi_i^T(\tau)\Gamma\xi_i(\tau) - \hat{\xi}_i^T(\tau)\Gamma\hat{\xi}_i(\tau))d\tau \\ &= -\check{d}_i(0) + \int_0^t (-\tilde{\xi}_i^T(\tau)\Gamma\tilde{\xi}_i(\tau) - 2\xi_i^T(\tau)\Gamma\tilde{\xi}_i(\tau))d\tau \\ &\le -\check{d}_i(0) + \frac{1}{3}\int_0^t \xi_i^T(\tau)\Gamma\xi_i(\tau)d\tau + 2\int_0^t \tilde{\xi}_i^T(\tau)\Gamma\tilde{\xi}_i(\tau)d\tau \\ &\le -\check{d}_i(0) + \frac{1}{3}\bar{d}_i(t) - \frac{1}{3}\bar{d}_i(0) + 4l_{ii}^2\sum_{j=1}^N \frac{\gamma_j\varepsilon_j(0)}{k_j}.\end{aligned}$$

Thus, we can obtain that

$$\begin{aligned}-\frac{3}{2}\xi^T(\check{D}\otimes\Gamma)\xi &= \sum_{i=1}^N(-\frac{3}{2}\check{d}_i)\xi_i^T\Gamma\xi_i \\ &\le \sum_{i=1}^N\Big(\frac{1}{2}\bar{d}_i(t) - \frac{3}{2}(\check{d}_i(0) + \frac{1}{3}\bar{d}_i(0)) \\ &\quad + 6\sum_{i=1}^N l_{ii}^2\sum_{j=1}^N \frac{\gamma_i\varepsilon_j(0)}{k_j}\Big)\xi_i^T\Gamma\xi_i \\ &\le \frac{1}{2}\xi^T(\bar{D}\otimes\Gamma)\xi + \xi^T(\beta_2 I_N\otimes\Gamma)\xi,\end{aligned} \tag{4.14}$$

where $\beta_2 \ge \max\{-\frac{3}{2}(\check{d}_i(0) + \frac{1}{3}\bar{d}_i(0)) + 6\sum_{i=1}^N l_{ii}^2\sum_{j=1}^N \frac{\gamma_i\varepsilon_j(0)}{k_j}\}$ is a positive constant.

(3) According to the triggering rule, we have

$$\begin{aligned}-2\xi^T(\tilde{D}\otimes\Gamma)\hat{\xi} &= -2\sum_{i=1}^N \tilde{d}_i\xi_i^T\Gamma\hat{\xi}_i \\ &\le \theta_1\sum_{i=1}^N \xi_i^T\Gamma\xi_i + \theta_1\sum_{i=1}^N \hat{\xi}_i^T\Gamma\hat{\xi}_i \\ &\le 3\theta_1\sum_{i=1}^N \xi_i^T\Gamma\xi_i + 2\theta_1\sum_{i=1}^N \tilde{\xi}_i^T\Gamma\tilde{\xi}_i \\ &\le 3\theta_1\sum_{i=1}^N \xi_i^T\Gamma\xi_i + 4\theta_1\sum_{i=1}^N l_{ii}^2\sum_{j=1}^N \tilde{x}_j^T\Gamma\tilde{x}_j.\end{aligned} \tag{4.15}$$

By substituting (4.13), (4.14), and (4.15) into (4.9), we can obtain that

$$\begin{aligned}\dot{V}_1 \leq& x^T[\mathcal{L}\otimes(AQ+A^TQ)]x+\xi^T[(\beta_2+3\theta_1)I_N\otimes\Gamma]\xi\\&+\sum_{i=1}^N(-\alpha+\beta V_0)\xi_i^T\Gamma\xi_i-\sum_{i=1}^N\beta\bar{d}_i\sum_{j=1}^N(\varepsilon_j+\frac{\sigma_j}{k_j}\tilde{x}_j^T\Gamma\tilde{x}_j)\\&+\sum_{i=1}^N(8\bar{d}_i+2\beta_1+4\theta_1)l_{ii}^2\sum_{j=1}^N\tilde{x}_j^T\Gamma\tilde{x}_j.\end{aligned}\tag{4.16}$$

By choosing $\beta \geq 2(4+\beta_1+2\theta_1)\max\{l_{ii}^2\}\max\{\frac{k_i}{\sigma_i}\}$, $\alpha=\alpha_1+\beta V_0(0)+\beta_2+3\theta_1$, we can obtain from (4.16) that

$$\begin{aligned}\dot{V}_1 &\leq x^T[\mathcal{L}\otimes(QA+A^TQ)]x-\xi^T(\alpha_1 I_N\otimes\Gamma)\xi\\&\leq x^T[\mathcal{L}\otimes(QA+A^TQ-\Gamma)]x\\&\leq\frac{1}{\lambda_N(\mathcal{L})}\xi^T[I_N\otimes(QA+A^TQ-\Gamma)]\xi\leq 0,\end{aligned}\tag{4.17}$$

where the second inequality is obtained by choosing $\alpha_1 \geq \frac{1}{\lambda_2(\mathcal{L})}$, and the last inequality is derived by the LMI (4.6). Note that $\lambda_2(\mathcal{L})$ and $\lambda_N(\mathcal{L})$ denote the smallest and largest nonzero eigenvalues of $\mathcal{L}$, respectively.

Therefore, $V_1(t)$ is bounded, and so are ξ_i and $\bar{d}_i$. In addition, V_1 has a finite limit as $t\to\infty$, denoted by V_1^∞. Integrating the third inequality of (4.17) yields

$$\int_0^\infty \xi^T[I_N\otimes(-(QA+A^TQ-\Gamma))]\xi dt\leq\lambda_N(\mathcal{L})(V_1(0)-V_1^\infty).$$

On the other hand, by (4.12) we can get that $d_i=\check{d}_i+\bar{d}_i\leq 2\bar{d}_i-\bar{d}_i(0)+d_i(0)+4l_{ii}^2\sum_{j=1}^N\frac{\gamma_j\varepsilon_j(0)}{k_j}$, implying that d_i is bounded. Then, $\check{d}_i=d_i-\bar{d}_i$ is also bounded. According to the triggering rule, $\tilde{x}_i^T\Gamma\tilde{x}_i$ is bounded, which implies that $(I_N\otimes BK)\tilde{x}$ is bounded. Following this, we obtain that $(I_N\otimes BK)\tilde{\xi}$ is bounded, which further implies that $(I_N\otimes BK)\hat{\xi}=(I_N\otimes BK)(\xi+\tilde{\xi})$ is also bounded. According to (4.5), we have

$$\begin{aligned}\dot{\xi}=&(I_N\otimes A)\xi+(\mathcal{L}\bar{D}\otimes BK)\xi+(\mathcal{L}D\otimes BK)\tilde{\xi}\\&+(\mathcal{L}\check{D}\otimes BK)\xi+(\mathcal{L}\tilde{D}\otimes BK)\hat{\xi},\end{aligned}\tag{4.18}$$

which is bounded. By Lemma 2.4, we can conclude that $\xi^T[I_N\otimes(-(QA+A^TQ-\Gamma))]\xi\to 0$ as $t\to\infty$. In this sense, $\xi\to 0$ as $t\to\infty$, i.e., consensus is achieved.

□

Remark 4.2 Note that to avoid continuous control updating, we use discrete value $\hat{d}_i(t)$ to design the control input $u_i(t)$. Due to its discontinuity, $\hat{d}_i(t)$ cannot be directly used to establish the Lyapunov function. To circumvent this difficulty, the virtual adaptive gains $\bar{d}_i$ are introduced to design the Lyapunov function V_1 given in (4.7). Furthermore, we also introduce the function V_0 into the Lyapunov function V_1, which is in certain sense similar to the idea of [63, 66], and the boundedness of V_0 facilitates the proof of Theorem 4.1.

The following result shows that Zeno behavior does not happen.

Theorem 4.2 *Under the conditions in Theorem 4.1, the Zeno behavior is excluded.*

Proof We rule out the Zeno behavior by providing a contradiction argument. Without loss of generality, suppose that agent i exists Zeno behavior, i.e., there exists $T < +\infty$ such that $\lim_{k\to\infty} t_k^i = T$. Then, for any positive constant Δ, there exists a positive integer k_0 such that for $k \geq k_0$, $t_k^i \in [T - \Delta, T)$.

According to Theorem 4.1, it is not difficult to find that x_i, $\hat{\xi}_i$, and $\hat{d}_i$ are bounded for time during $[0, T)$. Noting that $\tilde{x}_i(t_{k_0}^i) = 0$ and $\tilde{d}_i(t_{k_0}^i) = 0$, we have

$$
\begin{aligned}
\tilde{x}_i^T(t)\Gamma\tilde{x}_i(t) &= \left[\int_{t_{k_0}^i}^t Ax_i + BK\hat{d}_i\hat{\xi}_i(\tau)d\tau\right]^T \Gamma \left[\int_{t_{k_0}^i}^t Ax_i + BK\hat{d}_i\hat{\xi}_i(\tau)d\tau\right] \\
&\leq \|\Gamma\|h_1^2(t - t_{k_0}^i)^2, \\
|\tilde{d}_i(t)| &= \left|\int_{t_{k_0}^i}^t \hat{\xi}_i^T(\tau)\Gamma\hat{\xi}_i(\tau)d\tau\right| \leq h_2(t - t_{k_0}^i),
\end{aligned}
\tag{4.19}
$$

where h_1 and h_2 denote the upper bounds of $\|Ax_i(t) + BK\hat{d}_i(t)\hat{\xi}_i(t)\|$ and $\hat{\xi}_i^T(t)\Gamma\hat{\xi}_i(t)$ for $t \in [0, T]$, respectively.

Take $\Delta \triangleq \min\{\Delta_1, \Delta_2\}$, where $\Delta_1 = \sqrt{\frac{\gamma_i e^{-(k_i+\sigma_i\gamma_i)T}}{\|\Gamma\|h_1^2}}$ and $\Delta_2 = \frac{\theta_1 e^{-\theta_2 T}}{h_2}$. According to the triggering rule (4.3), the next event for agent i after $t_{k_0}^i$ happens if

$$
\tilde{x}_i(t)\Gamma\tilde{x}_i(t) \geq \gamma_i e^{-(k_i+\sigma_i\gamma_i)T} \vee |\tilde{d}_i(t)| \geq \theta_1 e^{-\theta_2 T}. \tag{4.20}
$$

Then, we can obtain that $t_{k_0+1}^i - t_{k_0}^i \geq \Delta$. Thus, we have $t_{k_0+1}^i \geq T$, which is inconsistent with $t_{k_0+1}^i < T$. Therefore, we can conclude that there does not exist the Zeno behavior. □

Remark 4.3 Theorems 4.1 and 4.2 show that the presented event-triggered protocol (4.2)–(4.4) solves the FDECDCC problem and excludes Zeno behavior. Different from the adaptive event-triggered protocols in [43, 58, 59, 81], here we use the law (4.3) with dynamic triggering functions (4.4) to determine the event instants. The advantage of the proposed fully distributed event-triggered protocol lies in that it only requires discrete communications and control updating. As a comparison, continuous updating of controllers cannot be avoided in [58, 59, 81], continuous communication cannot be avoided in [43], and the event-triggered protocols in [108] require global information. Different from the edge-based adaptive protocols given in Sect. 3.4 where the adaptive gains are designed for each edge, in this section we propose a node-based adaptive event-triggered protocol.

4.3 Fully Distributed Event-Based Control for Strongly Connected Graphs

In this section, we will further extend the adaptive event-triggered protocol (4.2) to the case of general directed graphs. It should be pointed out that the Laplacian matrix associated with directed graphs is asymmetric, which makes the Lyapunov function (4.7) no longer be positive-definite. For this reason, the protocols designed for undirected graphs are not applicable and novel protocols need to be explored in order to solve the FDECDCC problem. In the following, we first consider the case of strongly connected graphs. The key here is to tackle the difficulty caused by the unpleasant asymmetric Laplacian matrix.

An extra quadratic function is introduced for each agent and the controllers are designed as follows:

$$\begin{aligned} u_i(t) &= (\hat{d}_i(t) + \rho_i(t))K\hat{\xi}_i(t), \\ \dot{d}_i(t) &= \hat{\xi}_i(t)^T\Gamma\hat{\xi}_i(t), \end{aligned} \tag{4.21}$$

where $\rho_i(t) = \hat{\xi}_i^T(t)Q\hat{\xi}_i(t)$ is the quadratic form of $\hat{\xi}_i(t)$, $d_i(t)$ is the adaptive gain with initial value $d_i(0) \geq 1$, $\hat{d}_i(t) = d_i(t_k^i)$, and the rest variables are the same as those in (4.2).

To determine the triggering instants, we design the triggering condition for each agent as follows:

$$t_{k+1}^i = \inf\{t > t_k^i \mid \tilde{x}_i^T(t)\tilde{x}_i(t) \geq \gamma_i\varepsilon_i(t) \vee |\tilde{d}_i| \geq \theta_1 e^{-\theta_2(t-t_k^i)}\}, \tag{4.22}$$

where $t_0^i \triangleq 0$, $\gamma_i > 0$ is a positive constant, and $\varepsilon_i(t)$ is defined as in (4.4).

Denote $\bar{d}_i$ as the virtual adaptive gain with initial value $\bar{d}_i(0) \geq 1$, whose dynamics are given by

$$\dot{\bar{d}}_i = \xi_i^T\Gamma\xi_i.$$

Let $\check{d}_i = d_i - \bar{d}_i$, $\bar{\rho}_i = \xi_i^T Q \xi_i$, and $\check{\rho}_i = \rho_i - \bar{\rho}_i$. The closed-loop dynamics of x_i can be given by

$$\dot{x}_i = Ax_i + (\bar{d}_i + \bar{\rho}_i)BK\xi_i + (d_i + \rho_i)BK\tilde{\xi}_i + (\check{d}_i + \check{\rho}_i)BK\xi_i + \tilde{d}_i BK\hat{\xi}_i.$$

Then, the dynamics of the consensus error ξ in compact form can be written as

$$\begin{aligned} \dot{\xi} &= [I_N \otimes A + \mathcal{L}(\bar{D} + \bar{\rho}) \otimes BK]\xi + [\mathcal{L}(D + \rho) \otimes BK]\tilde{\xi} \\ &\quad + [\mathcal{L}(\check{D} + \check{\rho}) \otimes BK]\xi + [\mathcal{L}\tilde{D} \otimes BK]\hat{\xi}, \end{aligned} \tag{4.23}$$

where $\rho = \text{diag}\{\rho_1, \cdots, \rho_N\}$, $\bar{\rho} = \text{diag}\{\bar{\rho}_1, \cdots, \bar{\rho}_N\}$, and $\check{\rho} = \text{diag}\{\check{\rho}_1, \cdots, \check{\rho}_N\}$.

Next, we are going to show the achievement of consensus and the exclusion of Zeno behavior under strongly connected graphs.

Theorem 4.3 *Suppose that the network topology associated with the N agents in (4.1) is strongly connected. Consensus can be achieved under the adaptive event-triggered protocol (4.21) and (4.22) with gain matrices chosen as in Theorem 4.1. Moreover, the Zeno behavior is excluded.*

Proof Consider the following Lyapunov function candidate:

$$V_2 = \sum_{i=1}^{N} r_i \left[(\bar{d}_i + \frac{1}{2}\bar{\rho}_i)\bar{\rho}_i + \frac{(\bar{d}_i - \alpha)^2}{2} + \beta \bar{d}_i^2 V_0 \right], \tag{4.24}$$

where α and β are positive constants, r_i is defined by Lemma 2.3, and V_0 is defined in (4.8). The time derivative of V_2 is

$$\begin{aligned} \dot{V}_2 &= \xi^T[(\bar{D} + \bar{\rho})R \otimes (QA + A^T Q) - (\bar{D} + \bar{\rho})(R\mathcal{L} + \mathcal{L}^T R)(\bar{D} + \bar{\rho}) \otimes \Gamma]\xi \\ &\quad - 2\xi^T[(\bar{D} + \bar{\rho})R\mathcal{L}(D + \rho) \otimes \Gamma]\tilde{\xi} - 2\xi^T[(\bar{D} + \bar{\rho})R\mathcal{L}(\check{D} + \check{\rho}) \otimes \Gamma]\xi \\ &\quad - 2\xi^T[(\bar{D} + \bar{\rho})R\mathcal{L}\tilde{D} \otimes \Gamma]\hat{\xi} + \xi^T[((1 + 2\beta V_0)\bar{D} + \bar{\rho} - \alpha I_N)R \otimes \Gamma]\xi \\ &\quad - \sum_{i=1}^{N} \beta r_i \bar{d}_i^2 \sum_{j=1}^{N} \left(\varepsilon_j + \frac{\sigma_j}{k_j} \tilde{x}_j^T \tilde{x}_j \right). \end{aligned} \tag{4.25}$$

In light of Lemma 2.3, and noticing that

$$((\bar{D} + \bar{\rho})^{-1} r^T \otimes \mathbf{1}_p)^T ((\bar{D} + \bar{\rho}) \otimes B^T P)\xi = (r\mathcal{L} \otimes \mathbf{1}_p^T B^T P)x = 0,$$

we have

$$-\xi^T[(\bar{D}+\bar{\rho})(R\mathcal{L}+\mathcal{L}^T R)(\bar{D}+\bar{\rho})\otimes\Gamma]\xi \le -\xi^T\left[\frac{\lambda_2(\hat{L})}{N}(\bar{D}+\bar{\rho})^2\otimes\Gamma\right]\xi. \tag{4.26}$$

(1) We deal with the term $-2\xi^T[(\bar{D}+\bar{\rho})R\mathcal{L}(\check{D}+\check{\rho})\otimes\Gamma]\xi$. Similar to (4.11), we have

$$\tilde{\xi}_i^T\tilde{\xi}_i \le 2l_{ii}^2\sum_{j=1}^{N}\tilde{x}_j^T\tilde{x}_j \le 2l_{ii}^2\sum_{j=1}^{N}\gamma_j\varepsilon_j.$$

By using Lemma 2.5, we can obtain that

$$\begin{aligned}
|\check{\rho}_i| &= |\hat{\xi}_i^T Q\hat{\xi}_i - \xi_i^T Q\xi_i| = |\tilde{\xi}_i^T Q\tilde{\xi}_i + 2\tilde{\xi}_i^T Q\xi_i| \\
&\le \alpha_1\xi_i^T Q\xi_i + (1+\frac{1}{\alpha_1})\tilde{\xi}_i^T Q\tilde{\xi}_i \\
&\le \alpha_1\bar{\rho}_i + \lambda_{\max}(Q)(1+\frac{1}{\alpha_1})\tilde{\xi}_i^T\tilde{\xi}_i \\
&\le \alpha_1\bar{\rho}_i + 2\lambda_{\max}(Q)(1+\frac{1}{\alpha_1})l_{ii}^2\sum_{j=1}^{N}\gamma_j\varepsilon_j(0),
\end{aligned}$$

and

$$\begin{aligned}
|\check{d}_i| &= \left|\check{d}_i(0) + \int_0^t(\tilde{\xi}_i^T\Gamma\tilde{\xi}_i + 2\tilde{\xi}_i^T\Gamma\xi_i)d\tau\right| \\
&\le |\check{d}_i(0)| + \int_0^t(\alpha_1\xi_i^T\Gamma\xi_i + (1+\frac{1}{\alpha_1})\tilde{\xi}_i^T\Gamma\tilde{\xi}_i)d\tau \\
&\le |\check{d}_i(0)| - \alpha_1\bar{d}_i(0) + \alpha_1\bar{d}_i + \lambda_{\max}(\Gamma)(1+\frac{1}{\alpha_1})\int_0^t\tilde{\xi}_i^T\tilde{\xi}_i d\tau \\
&\le |\check{d}_i(0)| - \alpha_1\bar{d}_i(0) + \alpha_1\bar{d}_i + 2\lambda_{\max}(\Gamma)(1+\frac{1}{\alpha_1})l_{ii}^2\sum_{j=1}^{N}\frac{\gamma_j\varepsilon_j(0)}{k_j},
\end{aligned}$$

where $\alpha_1 = \frac{\lambda_2(\hat{L})}{4N\sigma_{\max}(R\mathcal{L})}$. Then, we have

$$|\check{d}_i + \check{\rho}_i| \le \alpha_1(\bar{d}_i + \bar{\rho}_i) + \beta_1,$$

where $\beta_1 = \max\{|\check{d}_i(0)| - \alpha_1\bar{d}_i(0) + 2(1+\frac{1}{\alpha_1})l_{ii}^2 \sum_{j=1}^{N}\left(\lambda_{\max}(Q) + \frac{\lambda_{\max}(\Gamma)}{k_j}\right)$ $\gamma_j\varepsilon_j(0)\}$. It follows that

$$
\begin{aligned}
-2\xi^T[(\bar{D}+\bar{\rho})R\mathcal{L}(\check{D}+\check{\rho})\otimes\Gamma]\xi \le \frac{\lambda_2(\hat{L})}{4N}\xi^T[(\bar{D}+\bar{\rho})^2\otimes\Gamma]\xi \\
+\frac{\sigma_{\max}(R\mathcal{L})}{\alpha_1}\xi^T[(\check{D}+\check{\rho})^2\otimes\Gamma]\xi,
\end{aligned}
\tag{4.27}
$$

where

$$
\begin{aligned}
\frac{\sigma_{\max}(R\mathcal{L})}{\alpha_1}\xi^T[(\check{D}+\check{\rho})^2\otimes\Gamma]\xi &= \frac{\sigma_{\max}(R\mathcal{L})}{\alpha_1}\sum_{i=1}^{N}(\check{d}_i+\check{\rho}_i)^2\xi_i^T\Gamma\xi_i \\
&\le \frac{\sigma_{\max}(R\mathcal{L})}{\alpha_1}\sum_{i=1}^{N}[\alpha_1(\bar{d}_i+\bar{\rho}_i)+\beta_1]^2\xi_i^T\Gamma\xi_i \\
&= \frac{\lambda_2(\hat{\mathcal{L}})}{4N}\xi^T[(\bar{D}+\bar{\rho})^2\otimes\Gamma]\xi \\
&\quad + 2\beta_1\sigma_{\max}(R\mathcal{L})\xi^T[(\bar{D}+\bar{\rho})\otimes\Gamma]\xi \\
&\quad + \frac{\beta_1^2\sigma_{\max}(R\mathcal{L})}{\alpha_1}\xi^T(I_N\otimes\Gamma)\xi.
\end{aligned}
\tag{4.28}
$$

Substituting (4.28) into (4.27) directly yields

$$
\begin{aligned}
-2\xi^T[(\bar{D}+\bar{\rho})R\mathcal{L}(\check{D}+\check{\rho})\otimes\Gamma]\xi \le \frac{\lambda_2(\hat{\mathcal{L}})}{2N}\xi^T[(\bar{D}+\bar{\rho})^2\otimes\Gamma]\xi \\
+ 2\beta_1\sigma_{\max}(R\mathcal{L})\xi^T[(\bar{D}+\bar{\rho})\otimes\Gamma]\xi \\
+ \frac{\beta_1^2\sigma_{\max}(R\mathcal{L})}{\alpha_1}\xi^T(I_N\otimes\Gamma)\xi.
\end{aligned}
\tag{4.29}
$$

(2) We then deal with the term $-2\xi^T[(\bar{D}+\bar{\rho})R\mathcal{L}(D+\rho)\otimes\Gamma]\tilde{\xi}$. Obviously, we have

$$\begin{aligned}\rho_i &= \hat{\xi}_i^T Q\hat{\xi}_i = \tilde{\xi}_i^T Q\tilde{\xi}_i + 2\tilde{\xi}_i^T Q\xi_i + \xi_i^T Q\xi_i \\ &\le (1+\alpha_1)\xi_i^T Q\xi_i + (1+\frac{1}{\alpha_1})\tilde{\xi}_i^T Q\tilde{\xi}_i \\ &\le (1+\alpha_1)\bar{\rho}_i + 2\lambda_{\max}(Q)(1+\frac{1}{\alpha_1})l_{ii}^2\sum_{j=1}^N \gamma_j\varepsilon_j(0),\end{aligned}$$

and

$$\begin{aligned}d_i &= d_i(0) + \int_0^t (\tilde{\xi}_i^T\Gamma\tilde{\xi}_i + 2\tilde{\xi}_i^T\Gamma\xi_i + \xi_i^T\Gamma\xi_i)d\tau \\ &\le d_i(0) - (1+\alpha_1)\bar{d}_i(0) + (1+\alpha_1)\bar{d}_i + (1+\frac{1}{\alpha_1})\int_0^t \tilde{\xi}_i^T\Gamma\tilde{\xi}_i d\tau \\ &\le d_i(0) - (1+\alpha_1)\bar{d}_i(0) + (1+\alpha_1)\bar{d}_i \\ &\quad + 2\lambda_{\max}(\Gamma)(1+\frac{1}{\alpha_1})l_{ii}^2\sum_{j=1}^N \frac{\gamma_j\varepsilon_j(0)}{k_j}.\end{aligned}$$

Then, we have

$$d_i + \rho_i \le (1+\alpha_1)(\bar{d}_i + \rho_i) + \beta_2.$$

It is clear to see that

$$\begin{aligned}-2\xi^T[(\bar{D}+\bar{\rho})R\mathcal{L}(D+\rho)\otimes\Gamma]\tilde{\xi} \le &\frac{\lambda_2(\hat{\mathcal{L}})}{4N}\xi^T[(\bar{D}+\bar{\rho})^2\otimes\Gamma]\xi \\ &+ \frac{\sigma_{\max}(R\mathcal{L})}{\alpha_1}\tilde{\xi}^T[(D+\rho)^2\otimes\Gamma]\tilde{\xi},\end{aligned}$$

where

$$\begin{aligned}\frac{\sigma_{\max}(R\mathcal{L})}{\alpha_1}\tilde{\xi}^T[(D+\rho)^2\otimes\Gamma]\tilde{\xi} &= \frac{\sigma_{\max}(R\mathcal{L})}{\alpha_1}\sum_{i=1}^N (d_i+\rho_i)^2\tilde{\xi}_i^T\Gamma\tilde{\xi}_i \\ \le \frac{\sigma_{\max}(R\mathcal{L})}{\alpha_1}\sum_{i=1}^N [(1+\alpha_1)(\bar{d}_i+\bar{\rho}_i)+\beta_2]^2\tilde{\xi}_i^T\Gamma\tilde{\xi}_i &\end{aligned}$$

$$= \frac{\sigma_{\max}(R\mathcal{L})}{\alpha_1} \sum_{i=1}^{N} [(1+\alpha_1)^2 \bar{d}_i^2 + 2(1+\alpha_1)\beta_2 \bar{d}_i + \beta_2^2] \tilde{\xi}_i^T \Gamma \tilde{\xi}_i$$

$$+ \frac{\sigma_{\max}(R\mathcal{L})}{\alpha_1} \sum_{i=1}^{N} [(1+\alpha_1)^2 \bar{\rho}_i + 2(1+\alpha_1)^2 \bar{d}_i + 2\beta_2(1+\alpha_1)] \bar{\rho}_i \tilde{\xi}_i^T \Gamma \tilde{\xi}_i.$$

Since $\bar{d}_i(t) \geq 1$, we can derive that

$$\begin{aligned}
&[(1+\alpha_1)^2 \bar{d}_i^2 + 2(1+\alpha_1)\beta_2 \bar{d}_i + \beta_2^2] \tilde{\xi}_i^T \Gamma \tilde{\xi}_i \\
&\leq [(1+\alpha_1)^2 + 2(1+\alpha_1)\beta_2 + \beta_2^2] \bar{d}_i^2 \tilde{\xi}_i^T \Gamma \tilde{\xi}_i \\
&\leq 2(1+\alpha_1+\beta_2)^2 \lambda_{\max}(\Gamma) l_{ii}^2 \bar{d}_i^2 \sum_{j=1}^{N} \tilde{x}_j^T \tilde{x}_j,
\end{aligned}$$

and

$$\begin{aligned}
&[(1+\alpha_1)^2 \bar{\rho}_i + 2(1+\alpha_1)^2 \bar{d}_i + 2\beta_2(1+\alpha_1)] \bar{\rho}_i \tilde{\xi}_i^T \Gamma \tilde{\xi}_i \\
&\leq \max\{(1+\alpha_1)^2, 2(1+\alpha_1)(1+\alpha_1+\beta_2)\} (\bar{d}_i + \bar{\rho}_i) \bar{\rho}_i \tilde{\xi}_i^T \Gamma \tilde{\xi}_i \\
&\leq \alpha_2 (\bar{d}_i + \bar{\rho}_i) \bar{\rho}_i \sum_{j=1}^{N} \gamma_j \varepsilon_j(t),
\end{aligned}$$

where $\alpha_2 = 2\max\{(1+\alpha_1)^2, 2(1+\alpha_1)(1+\alpha_1+\beta_2)\}\lambda_{\max}(\Gamma) l_{ii}^2$ and $\beta_2 = \max\left\{d_i(0) - (1+\alpha_1)\bar{d}_i(0) + 2(1+\frac{1}{\alpha_1}) l_{ii}^2 \sum_{j=1}^{N} \left(\lambda_{\max}(Q) + \frac{\lambda_{\max}(\Gamma)}{k_j}\right) \gamma_j \varepsilon_j(0)\right\}$. Since $\varepsilon_j(t) \leq \exp(-k_j t)\varepsilon_j(0)$, there exists a time instant T_0 such that

$$\alpha_2 \sum_{j=1}^{N} \gamma_j \varepsilon_j(t) \leq \frac{\alpha_1 \lambda_{\min}(R) \lambda_{\min}(X)}{2\sigma_{\max}(R\mathcal{L}) \lambda_{\max}(Q)},$$

holds for all $t > T_0$, where $X = -(QA + A^T Q - \Gamma) > 0$. Then, we obtain that for $t > T_0$,

$$\begin{aligned}
&\frac{\sigma_{\max}(RL)}{\alpha_1} \sum_{i=1}^{N} [(1+\alpha_1)^2 \bar{\rho}_i + 2(1+\alpha_1)^2 \bar{d}_i + 2\beta_2(1+\alpha_1)] \bar{\rho}_i \tilde{\xi}_i^T \Gamma \tilde{\xi}_i \\
&\quad \leq \frac{1}{2} \xi^T [(\bar{D} + \bar{\rho}) R \otimes X] \xi.
\end{aligned}$$

Thus, for $t > T_0$, we have

$$
\begin{aligned}
& -2\xi^T[(\bar{D}+\bar{\rho})R\mathcal{L}(D+\rho)\otimes\Gamma]\tilde{\xi} \\
& \le \frac{\lambda_2(\hat{\mathcal{L}})}{4N}\xi^T[(\bar{D}+\bar{\rho})^2\otimes\Gamma]\xi + \frac{1}{2}\xi^T[(\bar{D}+\bar{\rho})R\otimes X]\xi \\
& + \frac{\sigma_{\max}(R\mathcal{L})}{\alpha_1}\sum_{i=1}^{N}(1+\alpha_1+\beta_2)^2\bar{d}_i^2\tilde{\xi}_i^T\Gamma\tilde{\xi}_i.
\end{aligned} \tag{4.30}
$$

(3) According to the triggering rule, we can obtain that

$$
\begin{aligned}
& -2\xi^T[(\bar{D}+\bar{\rho})R\mathcal{L}\tilde{D}\otimes\Gamma]\hat{\xi} \\
& \le \frac{\lambda_2(\hat{\mathcal{L}})}{8N}\sigma_{\max}(R\mathcal{L})\xi^T[(\bar{D}+\bar{\rho})^2\otimes\Gamma]\xi + \frac{8N}{\lambda_2(\hat{\mathcal{L}})}\sigma_{\max}(R\mathcal{L})\hat{\xi}^T(\tilde{D}^2\otimes\Gamma)\hat{\xi} \\
& \le \frac{\lambda_2(\hat{\mathcal{L}})}{8N}\sigma_{\max}(R\mathcal{L})\xi^T[(\bar{D}+\bar{\rho})^2\otimes\Gamma]\xi + \frac{16N\theta_1^2}{\lambda_2(\hat{\mathcal{L}})}\sigma_{\max}(R\mathcal{L})\xi^T(I_N\otimes\Gamma)\xi \\
& + \frac{16N\theta_1^2}{\lambda_2(\hat{\mathcal{L}})}\sigma_{\max}(R\mathcal{L})\tilde{\xi}^T(I_N\otimes\Gamma)\tilde{\xi}.
\end{aligned} \tag{4.31}
$$

By substituting (4.29), (4.30), and (4.31) into (4.25), for $t > T_0$, by choosing $\alpha = \frac{\beta_1^2\sigma_{\max}(R\mathcal{L})}{\alpha_1} + \frac{16N\theta_1^2}{\lambda_2(\hat{\mathcal{L}})}\sigma_{\max}(R\mathcal{L}) + \alpha_0$ and $\beta \ge \left(\frac{2(1+\alpha_1+\beta_2)^2}{\alpha_1} + \frac{32N\theta_1^2}{\lambda_2(\hat{\mathcal{L}})}\right) \times \frac{\sigma_{\max}(R\mathcal{L})\lambda_{\max}(\Gamma)}{\lambda_{\min}(R)}\max\{l_{ii}^2\}\max\{\frac{k_i}{\sigma_i}\}$, we can get that

$$
\begin{aligned}
\dot{V}_2 \le\ & \xi^T[(\bar{D}+\bar{\rho})R\otimes(QA+A^TQ+\frac{1}{2}X+\beta_0\Gamma)]\xi \\
& -\xi^T[(\frac{\lambda_2(\hat{\mathcal{L}})}{8N}(\bar{D}+\bar{\rho})^2+\alpha_0 R)\otimes\Gamma]\xi,
\end{aligned} \tag{4.32}
$$

where $\beta_0 = 1 + 2\beta V_0(0) + \frac{2\beta_1\sigma_{\max}(R\mathcal{L})}{\lambda_{\min}(R)}$. By choosing $\alpha_0 \ge \frac{2N(1+\beta_0)^2\lambda_{\max}(R)}{\lambda_2(\hat{\mathcal{L}})}$ and using Lemma 2.5, we can derive that for $t > T_0$,

$$
\begin{aligned}
\dot{V}_2 \le\ & \xi^T[(\bar{D}+\bar{\rho})R\otimes(QA+A^TQ+\frac{1}{2}X+\beta_0\Gamma)]\xi \\
& -(1+\beta_0)\xi^T[(\bar{D}+\bar{\rho})R\otimes\Gamma]\xi \\
=\ & -\frac{1}{2}\xi^T[(\bar{D}+\bar{\rho})R\otimes X]\xi \le 0.
\end{aligned} \tag{4.33}
$$

Therefore, V_2 is bounded, and so are $\bar{d}_i$ and $\bar{\rho}_i$. Following similar steps in the proof of Theorem 4.1 and in light of Lemma 2.4, we can conclude that ξ_i asymptotically converges to zero, i.e., consensus is indeed realized. Similarly as in Theorem 4.2, we can derive that the Zeno behavior never happens.

□

Remark 4.4 It should be noted that in order to deal with the unpleasant asymmetric Laplacian matrix of directed graphs, we have included an extra quadratic function $\rho_i(t)$ into the designed controller of each agent. This idea is partly motivated by Lv et al. [64], which nevertheless considers the case of continuous control updating and continuous communication.

4.4 Fully Distributed Event-Based Control for Graphs Having a Directed Spanning Tree

In this section, we aim at considering general leader-follower directed graphs, satisfying the following assumption.

Assumption 4.1 The graph associated with the N agents has a directed spanning tree with the leader as a root.

Without loss of generality, we assume that the leader is indexed by 1 and $N-1$ followers are indexed by $2, \cdots, N$. Under this assumption, the Laplacian matrix $\mathcal{L}$ can be partitioned as $\mathcal{L} = \begin{bmatrix} 0 & 0_{1\times(N-1)} \\ \mathcal{L}_2 & \mathcal{L}_1 \end{bmatrix}$, where $\mathcal{L}_1$ is a non-singular M-matrix.

To track the leader's state, we propose a distributed adaptive event-based controller for each follower as

$$\begin{aligned} u_i(t) &= (\hat{d}_i(t) + \rho_i(t))K\hat{\xi}_i(t), \\ \dot{d}_i(t) &= \hat{\xi}_i^T(t)\Gamma\hat{\xi}_i(t),\ i = 2, \cdots, N, \end{aligned} \tag{4.34}$$

where $\rho_i = \hat{\xi}_i^T Q\hat{\xi}_i$, d_i is the adaptive gain with $d_i(0) \geq 1$, $\hat{d}_i(t) = d_i(t_k^i)$, and the rest variables are the same as those in (4.2).

Let $\tilde{x}_i = \hat{x}_i - x_i$ and $\tilde{d}_i = \hat{d}_i - d_i$. The triggering condition for each agent is designed by

$$\begin{aligned} t_{k+1}^1 &= \inf\{t > t_k^1 \mid \tilde{x}_1^T(t)\tilde{x}_1(t) \geq \gamma_1\varepsilon_1(t)\}, \\ t_{k+1}^i &= \inf\{t > t_k^i \mid \tilde{x}_i^T(t)\tilde{x}_i(t) \geq \gamma_i\varepsilon_i(t)\ \vee\ |\tilde{d}_i| \geq \theta_1 e^{-\theta_2(t-t_k^i)}\},\ i = 2, \cdots, N, \end{aligned} \tag{4.35}$$

where $t_0^i \triangleq 0$, $\gamma_i > 0$ is a positive constant, and $\varepsilon_i(t)$ is defined as in (4.4). At event instants t_k^1 of the leader, the leader sends its current information to the informed followers. When an event of a follower happens, the corresponding follower needs to update its controller and carry out communications with its neighboring followers.

Remark 4.5 For leader-follower networks, we have designed a triggering function for the leader to determine when to transmit information from the leader to informed followers that have direct access to the leader, and triggering functions for the followers to determine when to update their controllers and carry out communications with their neighboring followers.

Theorem 4.4 *Suppose that Assumption 4.1 holds. Under the adaptive event-triggered protocol composed of (4.34)–(4.35), the leader-follower consensus can be achieved and the time-varying weights $d_i(t)$ are bounded. Moreover, Zeno behavior does not exist.*

Proof Denote $\bar{d}_i$ as the virtual adaptive gain with initial value $\bar{d}_i(0) \geq 1$, $i = 2, \cdots, N$, whose dynamics are given by

$$\dot{\bar{d}}_i = \xi_i^T \Gamma \xi_i.$$

Let $\check{d}_i = d_i - \bar{d}_i$, $\bar{\rho}_i = \xi_i^T Q \xi_i$, and $\check{\rho}_i = \rho_i - \bar{\rho}_i$. The closed-loop dynamics of x_i can be given by

$$\begin{aligned} \dot{x}_i = {} & Ax_i + (\bar{d}_i + \bar{\rho}_i)BK\xi_i + (d_i + \rho_i)BK\tilde{\xi}_i + (\check{d}_i + \check{\rho}_i)BK\xi_i \\ & + (\tilde{d}_i + \tilde{\rho}_i)BK\hat{\xi}_i, \ i = 2, \cdots, N. \end{aligned} \tag{4.36}$$

Let $x_F = [x_2^T, \cdots, x_N^T]^T$, $\hat{x}_F = [\hat{x}_2^T, \cdots, \hat{x}_N^T]^T$, $\tilde{x}_F = [\tilde{x}_2^T, \cdots, \tilde{x}_N^T]^T$, $D_F = \text{diag}\{d_2, \cdots, d_N\}$, $\hat{D}_F = \text{diag}\{\hat{d}_2, \cdots, \hat{d}_N\}$, $\tilde{D}_F = \text{diag}\{\tilde{d}_2, \cdots, \tilde{d}_N\}$, $\rho_F = \text{diag}\{\rho_2, \cdots, \rho_N\}$, $\hat{\rho}_F = \text{diag}\{\hat{\rho}_2, \cdots, \hat{\rho}_N\}$, and $\xi_F = [\xi_2^T, \cdots, \xi_N^T]^T$. It is easy to verify that $\xi_F = (\mathcal{L}_1 \otimes I_n)(x_F - \mathbf{1} \otimes x_1)$. Since the leader-follower consensus problem is solved if and only if ξ_F asymptotically converges to zero, we refer to ξ_F as the consensus error. It satisfies the following dynamics:

$$\dot{\xi}_F = (I_{N-1} \otimes A)\xi_F + [\mathcal{L}_1(\hat{D}_F + \hat{\rho}_F) \otimes BK]\hat{\xi}_F. \tag{4.37}$$

Then, it is not difficult to rewrite (4.37) as

$$\begin{aligned} \dot{\xi}_F = {} & (I_{N-1} \otimes A)\xi_F + [\mathcal{L}_1(\bar{D}_F + \bar{\rho}_F) \otimes BK]\xi_F \\ & + [\mathcal{L}_1(D_F + \rho_F) \otimes BK]\tilde{\xi}_F + [\mathcal{L}_1\tilde{D}_F \otimes BK]\hat{\xi}_F \\ & + [\mathcal{L}_1(\check{D}_F + \check{\rho}_F) \otimes BK)]\xi_F. \end{aligned} \tag{4.38}$$

Consider the following Lyapunov function candidate:

$$V_3 = \sum_{i=2}^{N} s_i \left((\bar{d}_i + \frac{1}{2}\bar{\rho}_i)\bar{\rho}_i + \frac{(\bar{d}_i - \alpha)^2}{2} + \beta \bar{d}_i^2 V_0 \right), \tag{4.39}$$

where α and β are positive constants, s_i satisfies $S_F \triangleq \text{diag}\{s_2, \cdots, s_N\} > 0$ and $\hat{\mathcal{L}}_1 \triangleq S_F \mathcal{L}_1 + \mathcal{L}_1^T S_F > 0$, and $V_0 = \sum_{j=2}^{N} \frac{\varepsilon_j}{k_j}$. The time derivative of V_3 is

$$\begin{aligned}
\dot{V}_3 &= \sum_{i=2}^{N} s_i \left[(\bar{d}_i + \bar{\rho}_i)\dot{\bar{\rho}}_i + ((1 + 2\beta V_0)\bar{d}_i + \bar{\rho}_i - \alpha)\dot{\bar{d}}_i \right] + \sum_{i=2}^{N} \beta s_i \bar{d}_i^2 \dot{V}_0 \\
&= \xi_F^T [(\bar{D}_F + \bar{\rho}_F) S_F \otimes (QA + A^T Q) \\
&\quad - (\bar{D}_F + \bar{\rho}_F)(S_F \mathcal{L}_1 + \mathcal{L}_1^T S_F)(\bar{D}_F + \bar{\rho}_F) \otimes \Gamma] \xi_F \\
&\quad - 2\xi_F^T [(\bar{D}_F + \bar{\rho}_F) S_F \mathcal{L}_1 (D_F + \rho_F) \otimes \Gamma] \tilde{\xi}_F \\
&\quad - 2\xi_F^T [(\bar{D}_F + \bar{\rho}_F) S_F \mathcal{L}_1 (\check{D}_F + \check{\rho}_F) \otimes \Gamma] \xi_F \\
&\quad - 2\xi_F^T [(\bar{D}_F + \bar{\rho}_F) S_F \mathcal{L}_1 \tilde{D}_F \otimes \Gamma] \hat{\xi}_F \\
&\quad + \xi_F^T [((1 + 2\beta V_0)\bar{D}_F + \bar{\rho}_F - \alpha I_{N-1}) S_F \otimes \Gamma] \xi_F \\
&\quad - \sum_{i=2}^{N} \beta s_i \bar{d}_i^2 \sum_{j=1}^{N} \left(\varepsilon_j + \frac{\sigma_j}{k_j} \tilde{x}_j^T \tilde{x}_j \right).
\end{aligned} \tag{4.40}$$

Noting that $\hat{\mathcal{L}}_1 = S_F \mathcal{L}_1 + \mathcal{L}_1^T S_F > 0$, we have

$$\begin{aligned}
&- \xi_F^T [(\bar{D}_F + \bar{\rho}_F)(S_F \mathcal{L}_1 + \mathcal{L}_1^T S_F)(\bar{D}_F + \bar{\rho}_F) \otimes \Gamma] \xi_F \\
&\quad \le -\lambda_1(\hat{\mathcal{L}}_1) \xi_F^T [(\bar{D}_F + \bar{\rho}_F)^2 \otimes \Gamma] \xi_F,
\end{aligned} \tag{4.41}$$

where $\lambda_1(\hat{\mathcal{L}}_1)$ denotes the smallest eigenvalue of $\hat{\mathcal{L}}_1$.

A key in the rest of the proof is to deal with the three terms of (4.40), i.e., $-2\xi_F^T [(\bar{D}_F + \bar{\rho}_F) S_F \mathcal{L}_1 (D_F + \rho_F) \otimes \Gamma] \tilde{\xi}_F$, $-2\xi_F^T [(\bar{D}_F + \bar{\rho}_F) S_F \mathcal{L}_1 (\check{D}_F + \check{\rho}_F) \otimes \Gamma] \xi_F$, and $-2\xi_F^T [(\bar{D}_F + \bar{\rho}_F) S_F \mathcal{L}_1 \tilde{D}_F \otimes \Gamma] \hat{\xi}_F$.

Similar to (4.29), we can obtain that

$$\begin{aligned} & -2\xi_F^T[(\bar{D}_F+\bar{\rho}_F)S_F\mathcal{L}_1(\check{D}_F+\check{\rho}_F)\otimes\Gamma]\xi_F \\ \le & \frac{\lambda_1(\hat{\mathcal{L}}_1)}{2}\xi_F^T[(\bar{D}_F+\bar{\rho}_F)^2\otimes\Gamma]\xi_F \\ & +2\beta_1\sigma_{\max}(S_F\mathcal{L}_1)\xi_F^T[(\bar{D}_F+\bar{\rho}_F)\otimes\Gamma]\xi_F \\ & +\frac{\beta_1^2\sigma_{\max}(S_F\mathcal{L}_1)}{\alpha_1}\xi_F^T(I_{N-1}\otimes\Gamma)\xi_F, \end{aligned} \tag{4.42}$$

where $\alpha_1=\frac{\lambda_1(\hat{\mathcal{L}}_1)}{4\sigma_{\max}(S_F\mathcal{L}_1)}$ and

$$\beta_1=|\check{d}_i(0)|-\alpha_1\bar{d}_i(0)+2(1+\frac{1}{\alpha_1})l_{ii}^2\sum_{j=2}^N(\lambda_{\max}(Q)+\frac{\lambda_{\max}(\Gamma)}{k_j})\gamma_j\varepsilon_j(0).$$

Similar to (4.30), there exists T_0 such that for $\forall t>T_0$, it holds that

$$\begin{aligned} & -2\xi_F^T[(\bar{D}_F+\bar{\rho}_F)S_F\mathcal{L}_1(D_F+\rho_F)\otimes\Gamma]\tilde{\xi}_F \\ \le & \frac{\lambda_1(\hat{\mathcal{L}}_1)}{4}\xi_F^T[(\bar{D}_F+\bar{\rho}_F)^2\otimes\Gamma]\xi_F+\xi_F^T[(\bar{D}_F+\bar{\rho}_F)S_F\otimes X]\xi_F \\ & +\frac{\sigma_{\max}(S_F\mathcal{L}_1)}{\alpha_1}\sum_{i=2}^N(1+\alpha_1+\beta_2)^2\bar{d}_i^2\tilde{\xi}_i^T\Gamma\tilde{\xi}_i. \end{aligned} \tag{4.43}$$

Similar to (4.31), we have

$$\begin{aligned} & -2\xi_F^T[(\bar{D}_F+\bar{\rho}_F)S_F\mathcal{L}_1\tilde{D}_F\otimes\Gamma]\hat{\xi}_F \\ \le & \frac{\lambda_1(\hat{\mathcal{L}}_1)}{8}\sigma_{\max}(S_F\mathcal{L}_1)\xi_F^T[(\bar{D}_F+\bar{\rho}_F)^2\otimes\Gamma]\xi_F \\ & +\frac{16\theta_1^2}{\lambda_1(\hat{\mathcal{L}}_1)}\sigma_{\max}(S_F\mathcal{L}_1)\xi_F^T(I_{N-1}\otimes\Gamma)\xi_F \\ & +\frac{16\theta_1^2}{\lambda_1(\hat{\mathcal{L}}_1)}\sigma_{\max}(S_F\mathcal{L}_1)\tilde{\xi}_F^T(I_{N-1}\otimes\Gamma)\tilde{\xi}_F. \end{aligned} \tag{4.44}$$

According to (4.42)–(4.44), for $t>T_0$, by choosing

$$\alpha=\frac{\beta_1^2\sigma_{\max}(S_F\mathcal{L}_1)}{\alpha_1}+\frac{16\theta_1^2}{\lambda_1(\hat{\mathcal{L}}_1)}\sigma_{\max}(S_F\mathcal{L}_1)+\alpha_0,$$

and

$$\beta \geq \left(\frac{2(1+\alpha_1+\beta_2)^2}{\alpha_1} + \frac{32\theta_1^2}{\lambda_1(\hat{\mathcal{L}}_1)} \right) \frac{\sigma_{\max}(S_F \mathcal{L}_1)\lambda_{\max}(\Gamma)}{\lambda_{\min}(S_F)} \max\{l_{ii}^2\} \max\{\frac{k_i}{\sigma_i}\},$$

we can get that

$$\begin{aligned} \dot{V}_3 \leq\, & \xi_F^T[(\bar{D}_F + \bar{\rho}_F)S_F \otimes (QA + A^T Q + \frac{1}{2}X + \beta_0 \Gamma)]\xi \\ & - \xi^T[(\frac{\lambda_1(\hat{\mathcal{L}}_1)}{8}(\bar{D}_F + \bar{\rho}_F)^2 + \alpha_0 S_F) \otimes \Gamma]\xi_F, \end{aligned}$$

where $\beta_0 = 1 + 2\beta V_0(0) + \frac{2\beta_1 \sigma_{\max}(S_F \mathcal{L}_1)}{\lambda_{\min}(S_F)}$. By choosing $\alpha_0 \geq \frac{2(1+\beta_0)^2 \lambda_{\max}(S_F)}{\lambda_1(\hat{\mathcal{L}}_1)}$ and using Lemma 2.5, we can derive that for $t > T_0$,

$$\begin{aligned} \dot{V}_3 \leq\, & \xi_F^T[(\bar{D}_F + \bar{\rho}_F)S_F \otimes (QA + A^T Q + \frac{1}{2}X + \beta_0 \Gamma)]\xi_F \\ & - (1+\beta_0)\xi_F^T[(\bar{D}_F + \bar{\rho}_F)S_F \otimes \Gamma]\xi_F \\ = & -\frac{1}{2}\xi_F^T[(\bar{D}_F + \bar{\rho}_F)S_F \otimes X]\xi_F \leq 0. \end{aligned}$$

Therefore, V_3 is bounded, and so are $\bar{d}_i$ and $\bar{\rho}_i$. The rest is similar to the proofs of Theorems 4.1 and 4.2. The details are omitted for conciseness. □

Next, we use a numerical example to show the effectiveness of the proposed fully distributed event-triggered protocols. Specifically, we consider six agents whose dynamics satisfy (4.1) with

$$A = \begin{bmatrix} 0 & -1 & 0 \\ 1 & 0 & 0 \\ 0 & 0 & 0.1 \end{bmatrix}, \quad B = \begin{bmatrix} 0 \\ 1 \\ 1 \end{bmatrix}.$$

The agents are networked by the graph shown in Fig. 4.1, which has a directed spanning tree with agent 1 as a root.

To achieve the consensus of six agents, we use the adaptive event-triggered protocol composed of (4.34)–(4.35). Solving the LMI (4.6) gives

$$P = \begin{bmatrix} 0.4298 & 0.1428 & 0.2837 \\ 0.1428 & 0.7147 & 0.4281 \\ 0.2837 & 0.4281 & 0.7638 \end{bmatrix}.$$

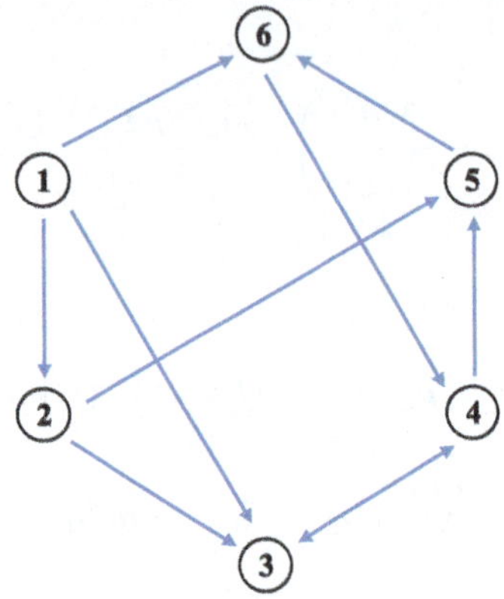

Fig. 4.1 The network topology associated with the agents

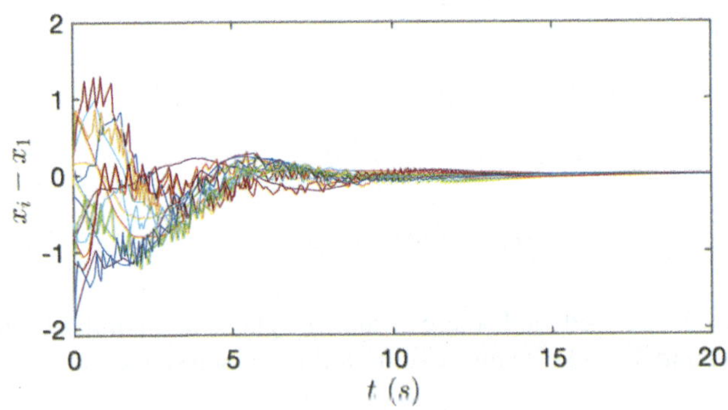

Fig. 4.2 The consensus errors $x_i(t) - x_1(t)$, $i = 2, \cdots, 6$

Then, we obtain the feedback matrices

$$K = \begin{bmatrix} 1.1004 & -0.8883 & -1.2200 \end{bmatrix}, \; \Gamma = \begin{bmatrix} 1.2109 & -0.9775 & -1.3425 \\ -0.9775 & 0.7891 & 1.0837 \\ -1.3425 & 1.0837 & 1.4884 \end{bmatrix}.$$

The rest parameters of the protocol are selected as $\gamma_i = 1.5$, $\theta_1 = 2$, $\theta_2 = 1.5$, $k_i = 0.3$, $\sigma_i = 0.2$, and $\varepsilon_i(0) = 0.3$, $i = 1, \cdots, 6$.

Running the software MATLAB in a computing frequency of 1 GHz, we can get the following results. Figure 4.2 shows that consensus can be achieved. The time-varying gains $d_i(t)$ are shown in Fig. 4.3, implying that $d_i(t)$ tend to fixed values. According to Fig. 4.4, each control input is updated only at event instants and is via zero-order hold until next event instant. Communications for each agent are only needed at the discrete instants shown as in Fig. 4.5.

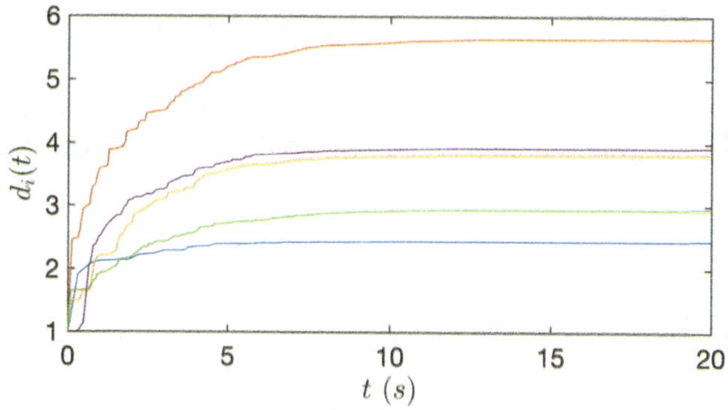

Fig. 4.3 The time-varying gains $d_i(t)$, $i = 2, \cdots, 6$

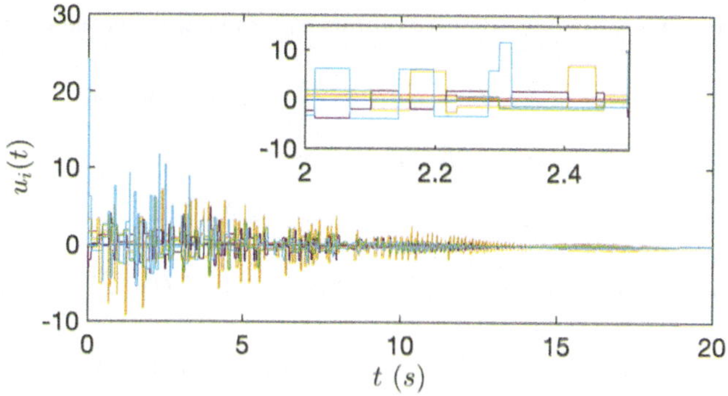

Fig. 4.4 Control inputs $u_i(t)$, $i = 2, \cdots, 6$

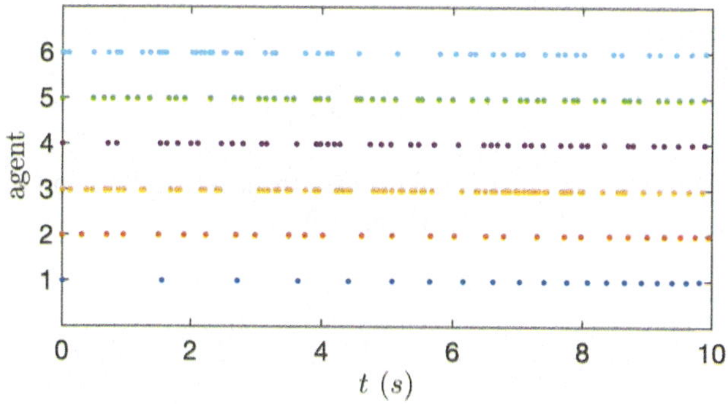

Fig. 4.5 Triggering instants of each agent

4.5 Output Feedback Event-Triggered Control

In this section, we are going to extend the above results to the case where the state information is not available. In this case, we need to design novel output feedback consensus protocols to solve the FDECDCC problem.

4.5.1 The Undirected Graph Case

We first consider the case of undirected graphs in this subsection. To deal with the difficulty caused by unavailable state information, the idea here is to design a Luenberger observer for each agent using local output information, based on which the controllers and triggering functions will be established.

We propose the following adaptive output-based event-triggered controller:

$$\begin{aligned}\dot{v}_i(t) &= Av_i(t) + Bu_i(t) + F(Cv_i(t) - y_i(t)),\\ \dot{e}_i(t) &= \hat{\eta}_i^T(t)\Gamma\hat{\eta}_i(t),\\ u_i(t) &= \hat{e}_i(t)K\hat{\eta}_i(t),\ i = 1, \cdots, N,\end{aligned} \tag{4.45}$$

where $v_i(t)$ are the observer's state, $e_i(0) \geq 1$, $\hat{e}_i(t) = e_i(t_k^i)$, $\hat{\eta}_i(t) = \sum_{j=1}^N a_{ij}(\hat{v}_i(t) - \hat{v}_j(t))$, and $\hat{v}_i = v_i(t_k^i)$, $\forall t \in [t_k^i, t_{k+1}^i)$, $i = 1, \cdots, N$.

By defining $\tilde{v}_i = \hat{v}_i - v_i$ and $\tilde{e}_i = \hat{e}_i - e_i$ as the estimation errors, we design the following triggering condition:

$$t_{k+1}^i \triangleq \inf\{t > t_k^i \mid \tilde{v}_i^T(t)\Gamma\tilde{v}_i(t) \geq \gamma_i\epsilon_i(t) \vee |\tilde{e}_i| \geq \theta_1 e^{-\theta_2(t-t_k^i)}\}, \tag{4.46}$$

where $\epsilon_i(t)$ satisfies

$$\dot{\epsilon}_i(t) = -k_i\epsilon_i(t) - \sigma_i\tilde{v}_i^T(t)\Gamma\tilde{v}_i(t), \tag{4.47}$$

with $\epsilon_i(0) > 0$ and the other parameters being the same as those in (4.3).

Let $z_i = v_i - x_i$, $i = 1, \cdots, N$. Then, we have

$$\begin{aligned}\dot{z}_i &= (A + FC)z_i,\\ \dot{v}_i &= Av_i + Bu_i + FCz_i,\ i = 1, \cdots, N.\end{aligned}$$

Denote $\eta = [\eta_1^T, \cdots, \eta_N^T]^T = (\mathcal{L} \otimes I_n)v$. Letting $\tilde{\eta}_i = \sum_{j=1}^N a_{ij}(\tilde{v}_i - \tilde{v}_j)$, we have $\tilde{\eta}_i = \hat{\eta}_i - \eta_i$. Denote $\bar{e}_i$ as the virtual adaptive gain, whose dynamics are given by

$$\dot{\bar{e}}_i = \eta_i^T\Gamma\eta_i,$$

with initial value $\bar{e}_i(0) \geq 1$. Further let $\check{e}_i = e_i - \bar{e}_i$. Then, we can derive that the closed-loop dynamics of v_i satisfy

$$\dot{v}_i = Av_i + \bar{e}_i BK\eta_i + e_i BK\tilde{\eta}_i + \check{e}_i BK\eta_i + \tilde{e}_i BK\hat{\eta}_i + FCz_i,$$

which can be written in a compact form as

$$\begin{aligned}\dot{v} = &(I_N \otimes A)v + (\bar{E} \otimes BK)\eta + (E \otimes BK)\tilde{\eta} + (\check{E} \otimes BK)\eta \\ &+ (\tilde{E} \otimes BK)\hat{\eta} + (I_N \otimes FC)z,\end{aligned} \tag{4.48}$$

where $\tilde{\eta} = [\tilde{\eta}_1^T, \cdots, \tilde{\eta}_N^T]^T$, $\hat{\eta} = [\hat{\eta}_1^T, \cdots, \hat{\eta}_N^T]^T$, $\bar{E} = \text{diag}\{\bar{e}_1, \cdots, \bar{e}_N\}$, $E = \text{diag}\{e_1, \cdots, e_N\}$, $\tilde{E} = \text{diag}\{\tilde{e}_1, \cdots, \tilde{e}_N\}$, $\check{E} = \text{diag}\{\check{e}_1, \cdots, \check{e}_N\}$, and $z = [z_1^T, \cdots, z_N^T]^T$.

Now, we present the main result under the proposed adaptive output feedback event-triggered protocol.

Theorem 4.5 *Choose F such that $A + FC$ is Hurwitz and choose $K = -B^TQ$ and $\Gamma = QBB^TQ$ where $Q > 0$ is the solution to the following ARE:*

$$A^TQ + QA - QBB^TQ + I = 0. \tag{4.49}$$

Then, the agents can achieve consensus and do not exist Zeno behavior.

Proof Consider the following Lyapunov function candidate:

$$V_3 = v^T(\mathcal{L} \otimes Q)v + \sum_{i=1}^{N}\left[\frac{1}{2}(\bar{e}_i - \alpha)^2 + \beta\bar{d}_i V_\epsilon + \gamma z_i^T Q_0 z_i\right], \tag{4.50}$$

where α, β, and γ are positive constants, $V_\epsilon = \sum_{j=1}^{N} \frac{\epsilon_j}{k_j}$, and Q_0 is a positive definite matrix such that $Q_0(A + FC) + (A + FC)^TQ_0 < 0$.

When computing the time derivative of V_3, we need to note that

$$2v^T(\mathcal{L} \otimes Q)(I_N \otimes FC)z \leq \frac{1}{2}v^T(\mathcal{L} \otimes I_n)v + 2\lambda_N(\mathcal{L})\lambda_{\max}(C^TF^TQ^2FC)z^Tz,$$

and

$$\begin{aligned}\frac{dz_i^TQ_0z_i}{dt} &= z_i^T[Q_0(A + FC) + (A + FC)^TQ_0]z_i \\ &\leq -\lambda_{\min}(Q_0(A + FC) + (A + FC)^TQ_0)z_i^Tz_i.\end{aligned}$$

By following the steps and choosing α and β as in the proof of Theorem 4.1, and letting $\gamma \geq \frac{2\lambda_N(\mathcal{L})\lambda_{\max}(C^T F^T Q^2 FC)}{\lambda_{\min}(Q_0(A+FC)+(A+FC)^T Q_0)}$, we can conclude that

$$\begin{aligned} \dot{V}_3 &\leq x^T[\mathcal{L} \otimes (QA + A^T Q - \Gamma + \frac{1}{2}I)]x \\ &\leq \frac{1}{\lambda_N(\mathcal{L})}\xi^T \left[I_N \otimes (QA + A^T Q - \Gamma + \frac{1}{2}I) \right] \xi \leq 0. \end{aligned} \tag{4.51}$$

Then, we can show that v_i asymptotically converge to be the same, implying that consensus can be achieved. The Zeno behavior can be excluded as in the proof of Theorem 4.2. The details are skipped for brevity. □

Remark 4.6 To avoid using the state information, in this subsection we propose a kind of adaptive observer-based event-triggered protocol (4.45). Note that the design of the observer is not unique and other kinds of observers may also work. To be convenient, we choose the design matrix Q by solving the ARE (4.49). Actually, the solvability of the Riccati equation (4.49) is equal to that of the LMI (4.6), i.e., (A, B) is stabilizable [53].

4.5.2 The Directed Graph Case

In this subsection, we consider strongly connected graphs and extend the results in Sect. 4.3 to the case where only local output information is available.

We propose the following adaptive output-based event-triggered controller for each agent:

$$\begin{aligned} \dot{v}_i(t) &= Av_i(t) + Bu_i(t) + F(Cv_i(t) - y_i(t)), \\ \dot{e}_i(t) &= \hat{\eta}_i^T(t)\Gamma\hat{\eta}_i(t), \\ u_i(t) &= (\hat{e}_i(t) + \varrho_i(t))K\hat{\eta}_i(t),\ i = 1, \cdots, N, \end{aligned} \tag{4.52}$$

where $e_i(0) \geq 1$, $\hat{e}_i(t) = e_i(t_k^i)$, $\hat{\eta}_i(t) = \sum_{j=1}^N a_{ij}(\hat{v}_i(t) - \hat{v}_j(t))$, $\hat{v}_i(t) = v_i(t_k^i)$, and $\varrho_i = \hat{\eta}_i^T Q\hat{\eta}_i$, $\forall t \in [t_k^i, t_{k+1}^i)$, $i = 1, \cdots, N$.

To determine the event instants, we design the following triggering condition:

$$t_{k+1}^i \triangleq \inf\{t > t_k^i \mid \tilde{v}_i^T(t)\tilde{v}_i(t) \geq \gamma_i\epsilon_i(t) \vee |\tilde{e}_i| \geq \theta_1 e^{-\theta_2(t-t_k^i)}\}, \tag{4.53}$$

where $\epsilon_i(t)$ is defined as in (4.47).

Theorem 4.6 *Choosing feedback matrices F, K, and Γ as in Theorem 4.5, the consensus of (4.1) can be achieved and the time-varying weights $e_i(t)$ are bounded. Moreover, there does not exist Zeno behavior.*

Theorem 4.6 can be proved by following the steps as in the proofs of Theorems 4.3 and 4.5. Here we omit the details for brevity. Under the adaptive output-feedback event-triggered protocol composed of (4.52) and (4.53), the observer v_i will asymptotically converge to the state x_i.

4.6 Fully Distributed Event-Triggered Control with Relative State Measurements

In Sects. 4.2–4.5, we have proposed fully distributed adaptive event-triggered protocols with both discrete communication and updating. The core is to design a dynamic triggering condition for each agent and when it is triggered, the agent sends its current state information to its out-neighbors. Naturally, there is a different idea that when the triggering condition is triggered, the agent samples information from its neighbors and updates its controller. Following this idea, the event-triggered protocols with relative state measurement information are studied in [58, 59], which, however, needs continuous control updating.

For illustration, in this section we consider the case of strongly connected directed graph. Motivated by Cheng et al. [18], Li et al. [59], and Li et al. [58], we propose a distributed adaptive event-based controller for each follower as

$$\begin{aligned} u_i(t) &= (c_i(t_k^i) + \rho_i(t_k^i))K\xi_i(t_k^i), \\ \dot{c}_i(t) &= \kappa_i \xi_i^T(t_k^i)\Gamma\xi_i(t_k^i), \\ \rho_i(t) &= \xi_i^T P\xi_i, \ i = 1, \cdots, N, \end{aligned} \tag{4.54}$$

where $\rho_i(t) = \xi_i^T(t)Q\xi_i(t)$, $c_i(0)$ and κ_i are positive constants, $Q > 0$, K, and Γ are constant matrices, $\xi_i(t) = \sum_{j=1}^N a_{ij}(x_i(t) - x_j(t))$, and t_k^i denotes the k-th event instant of agent i. Denote $\hat{c}_i(t) \triangleq c_i(t_k^i)$ and $\hat{\rho}_i(t) \triangleq \rho_i(t_k^i)$.

We define the following triggering functions:

$$\begin{aligned} f_{1i} &= (c_i + \rho_i)^2 \tilde{\xi}_i^T PBB^T P\tilde{\xi}_i - \phi_1 \xi_i^T PBB^T P\xi_i - \mu_{1i}(t), \\ f_{2i} &= (\tilde{c}_i + \tilde{\rho}_i)^2 \hat{\xi}_i^T PBB^T P\hat{\xi}_i - \phi_2 \xi_i^T PBB^T P\xi_i - \mu_{2i}(t), \ i = 1, \cdots, N, \end{aligned} \tag{4.55}$$

where $\tilde{c}_i = \hat{c}_i - c_i$, $\tilde{\rho}_i = \hat{\rho}_i - \rho_i$, $\hat{\xi}_i(t) = \xi_i(t_k^i)$, $\tilde{\xi}_i = \hat{\xi}_i - \xi_i$, ϕ_1 and ϕ_2 are positive constants, and $\mu_{1i}(t)$ and $\mu_{2i}(t)$ are positive scalar functions belonging to $L_1[0, +\infty)$, indicating that $L_1[0, +\infty) \triangleq \{x(t) \in \mathbf{R} | \int_0^\infty |x(\tau)| d\tau < \infty\}$. Exactly, $\mu e^{-\nu t}$ can be a special case of $\mu_{1i}(t)$ and $\mu_{2i}(t)$. We define $t_{k+1}^i \triangleq \inf\{t > t_k^i \mid f_{1i}(t) \geq 0 \vee f_{2i}(t) \geq 0\}$. Once $f_{1i} \geq 0$ or $f_{2i} \geq 0$, agent i immediately samples information from its neighbors and updates its control input $u_i(t)$, otherwise $u_i(t)$ keeps zero-order hold.

Then, the consensus error ξ satisfies the following dynamics:

$$\dot{\xi} = (I_N \otimes A)\xi + [\mathcal{L}(\hat{C} + \hat{\rho}) \otimes BK]\hat{\xi}, \tag{4.56}$$

where $\hat{C} = \mathrm{diag}\{\hat{c}_1^T, \cdots, \hat{c}_N^T\}$ and $\hat{\rho} = \mathrm{diag}\{\hat{\rho}_1^T, \cdots, \hat{\rho}_N^T\}$.

Theorem 4.7 *Suppose that the graph $\mathcal{G}$ is strongly connected. Letting $c_i(0) > \sqrt{2\phi_1}$, under the adaptive event-triggered protocol composed of (4.54) and (4.55) with $K = -B^T Q$ and $\Gamma = QBB^T Q$, where $Q > 0$ is the solution to the LMI (4.1), consensus can be achieved and Zeno behavior can be excluded.*

Theorem 4.7 can be proved similarly as in [18] and the details are omitted here for brevity. Note that the triggering function (4.55) relies on the continuous relative state information $\xi_i(t)$ of neighboring agents. Even if as a compensation we can compute $\xi_i(t)$ according to the dynamics $\dot{\xi}_i(t) = A\xi_i(t) + B\sum_{j=1}^{N} a_{ij}(u_i(t) - u_j(t))$, each agent needs to use the inputs of neighbors as pointed out in [16, 18, 81]. This means that each agent needs to transmit its input value to its out-neighbors at event instants. In addition, the computing of the external variable $\xi_i(t)$ also increases computational cost. Furthermore, the reliability of the calculated consensus error ξ_i depends on how accurate the dynamics are. Note that once there exist unmodeled dynamics or disturbances, $\xi_i(t)$ cannot be accurately calculated by the above computing equation anymore. These drawbacks essentially exist in the proposed protocol (4.54) as well as the existing protocols based on relative state measurements, such as those in [58, 59]. Nevertheless, such limitations can be circumvented by using the distributed adaptive dynamic event-triggered protocols proposed in Sect. 4.3.

4.7 Notes

The results of this chapter are mainly based on [18, 19]. There exist some related works, which also consider the fully distributed event-triggered consensus problem. For example, fully distributed consensus protocols under undirected graphs were presented in [81], whose design structure is, however, hard to solve the consensus problem constrained by the discrete control updating and directed graphs. Owing to the proposed event-triggering functions depending on only local information, the presented adaptive protocols in this chapter can well solve the FDECDCC problem, and thus may be of more general significance than those in [43, 81, 108].

Chapter 5
Fully Distributed Event-Triggered Affine Formation Control

In the previous chapters, fully distributed event-triggered protocols have been presented to achieve consensus for multi-agent systems. In practice, a multi-agent system usually needs to execute some tasks while keeping certain specific geometric patterns, which gives rise to the formation control problem. There are mainly three types of methods to design formation control algorithms, namely, the position-based, displacement-based, and distance-based ones [75]. Nevertheless, these methods have their respective limitations when considering the formation maneuverability. For example, it would be really intractable to control the formation scale or orientation by only resorting to the displacement-based control method [116]. To this end, the affine formation maneuver control, which aims to steer a group of agents to simultaneously execute various formation maneuvers (e.g., translations, scales, and rotations), has been put forward. Ever since the pioneer work in [61], considerable attention has been devoted to affine formation maneuver control issues for multi-agent systems with double-integrator, high-order, and general linear dynamics over undirected graphs [9, 78, 116]. For the case of directed graphs, the corresponding control issues have been investigated in [23, 60, 99] for double-integrator and general linear multi-agent systems. However, due to the asymmetry induced by directed graphs, the existing results usually require the knowledge of certain global information of the network and thus are not fully distributed. This is true especially for the case with event-triggered communication among agents. By following similar lines of fully distributed event-triggered consensus in Chaps. 3 and 4, we intend in this chapter to establish fully distributed event-triggered affine formation control schemes.

In Sect. 5.1, we first consider the static formation control problem for general linear multi-agent systems with event-triggered communication. Fully distributed event-based formation controllers are presented, by adapting the adaptive protocols in Chap. 4 with the geometrical information of the given static formation included.

In Sect. 5.2, we consider the event-triggered affine formation control problem for multi-agent systems composed of several leaders and followers over general

Z. Li et al., *Distributed Event-triggered Control*,
https://doi.org/10.1007/978-981-99-8170-0_5

directed graphs. In order to achieve affine formation, a stress matrix is utilized to represent the interactions among the agents. Note that the stress matrix can have both positive and negative non-diagonal elements and may have eigenvalues with negative real parts, which is essentially different from the standard Laplacian matrix. This poses additional difficulty into the task of designing fully distributed event-triggered affine formation control laws. One contribution of this section is that the in-degree information of each node is incorporated into the control law to avoid the need for global information of the network. In Sect. 5.3, the σ-modification technique is adopted to rule out the parameter drift phenomenon in the presence of external disturbances. Compared with the related existing works, e.g., [9, 23, 60, 78, 99, 103, 116], which either are applicable to undirected graphs or require global information, the proposed event-triggered affine formation control scheme is fully distributed and works for general linear multi-agent systems with directed graphs.

In Sect. 5.4, we provide some numerical simulations and hardware experiments to demonstrate the effectiveness of the proposed control schemes. The hardware experiments are carried out in an indoor laboratory, where a group of nano quadrotors called Crazyflie [5] achieves affine formation with good precision and the Optitrack, a motion capture system, provides the Crazyflies' position information.

5.1 Static Formation Control with Event-Triggered Communication

In this section, we will extend the adaptive event triggered consensus protocols proposed in the previous chapter to handle the static formation control problem. For simplicity, still consider a group of linear agents:

$$\dot{x}_i = Ax_i + Bu_i, \ i = 1, \cdots, N, \tag{5.1}$$

where $x_i \in \mathbf{R}^d$ and $u_i \in \mathbf{R}^p$ denote the state and the control input of agent i, respectively.

Let $\widetilde{H} \triangleq (h_1, \cdots, h_N) \in \mathbf{R}^{d \times N}$ describe a constant formation structure of a multi-agent system in a reference coordinate frame, where $h_i \in \mathbf{R}^d$ is the formation variable corresponding to agent i and satisfying $Ah_i = 0$. Then, variable $h_i - h_j$ can be used to denote the relative formation vector between agents i and j, which is independent of the reference coordinate.

The agents (5.1) are said to achieve a given static formation $\widetilde{H}$, if $\lim_{t\to\infty} \|(x_i(t) - h_i) - (x_j(t) - h_j)\| = 0, \forall i, j = 1, \cdots, N$.

Denote $\hat{x}_j(t) = x_j(t_k^j), \forall t \in [t_k^j, t_{k+1}^j)$ as the state estimation of agent j, where t_k^j denotes the k-th event-triggering instant of agent j. Note that the formation tracking error ξ can be defined as $\xi_i(t) = \sum_{j=1}^N a_{ij}(x_i(t) - h_i - x_j(t) + h_j)$.

By adopting the similar idea in Chap. 4 as well as the displacement-based formation scheme, we propose a distributed adaptive event-triggered formation control law as follows:

$$\begin{aligned} u_i(t) &= (d_i(t) + \rho_i(t))K\hat{\xi}_i(t), \\ \dot{d}_i(t) &= \hat{\xi}_i(t)^T \Gamma \hat{\xi}_i(t),\ i = 1, \cdots, N, \end{aligned} \tag{5.2}$$

where $\hat{\xi}_i(t) = \sum_{j=1}^{N} a_{ij}(\hat{x}_i(t) - h_i - \hat{x}_j(t) + h_j)$, $\rho_i(t) = \hat{\xi}_i^T(t)Q\hat{\xi}_i(t)$, and $d_i(t)$ is the adaptive gain with initial value $d_i(0) \geq 1$.

The triggering condition for each agent is defined as follows:

$$t_{k+1}^i = \inf\{t > t_k^i \mid \tilde{x}_i^T(t)\tilde{x}_i(t) \geq \gamma_i \varepsilon_i(t)\}, \tag{5.3}$$

where $\tilde{x}_i(t) = \hat{x}_i(t) - x_i(t)$ and $t_0^i \triangleq 0$, $\gamma_i > 0$ is a positive constant, and $\varepsilon_i(t)$ satisfies the following dynamics:

$$\dot{\varepsilon}_i(t) = -k_i\varepsilon_i(t) - \sigma_i\tilde{x}_i^T(t)\tilde{x}_i(t), \tag{5.4}$$

with $\varepsilon_i(0) > 0$, $k_i > 0$, and $\sigma_i > 0$.

Note that the control updating of u_i in (5.2) is continuous. Thus, the control law in (5.2) is built upon a simplified form of (4.21) which utilizes discrete control updating. The triggering function (5.3) is modified correspondingly.

The feedback gain matrices K and Γ and other parameters in (5.2) and (5.3) can be designed as in Chap. 4. By following similar steps in the proofs of Theorems 4.1 and 4.2, we can readily show that static formation can be achieved under the proposed event-triggered control law. The details are omitted here for conciseness.

5.2 Affine Formation Control with Event-Triggered Communication

In this section, we are going to consider the affine formation maneuver control problem of linear multi-agent systems with event-triggered communication.

5.2.1 Formulation of Affine Formation Problem

The agent dynamics are described by (5.1) in $\mathbf{R}^d$ with $d \geq 2$. Assume that there are M leaders with zero control inputs and $N - M$ followers in the network. Let $\mathcal{V}_f$ and $\mathcal{V}_l$ be the sets of followers and leaders, respectively. The communication graph among the N agents is denoted by a directed graph $\mathcal{G} = (\mathcal{V}, \mathcal{E})$. Apart from the basic definitions of graph theory in Chapter 2, we need to additionally introduce the notion of κ-reachability. A node v in $\mathcal{G}$ is set as κ-reachable ($\kappa \geq 2$) from a nonsingleton set $\mathcal{U}$ of nodes when there is a path from a node in $\mathcal{U}$ to v after deleting any $\kappa - 1$ nodes except v.

Let $r = [r_1^T, \cdots, r_N^T]^T \in \mathbf{R}^{dN}$ be the nominal configuration of the agents. A constant formation, denoted by $(\mathcal{G}, r)$, is the graph $\mathcal{G}$ with its vertex i mapped to point r_i. Then, the target formation can be described by

$$p^*(t) = [I_N \otimes D(t)]r + \mathbf{1}_N \otimes b(t), \tag{5.5}$$

where $D(t) \in \mathbf{R}^{d\times d}$, and $b(t) \in \mathbf{R}^d$ represents the formation's translational maneuver. Thus, the desired position of agent i in the target formation is

$$p_i^*(t) = D(t)r_i + b(t).$$

The target configuration is actually a time-varying affine transformation of the nominal configuration. The affine transformation, which may be translation, rotation, scale, shear or a combination of them, can be realized by choosing appropriate matrix $D(t)$ in (5.5).

For a formation $(\mathcal{G}, r)$, a stress is a set of scalars $\{\omega_{ij}\}_{(i,j)} \in \mathcal{E}$, assigned to all the edges. The value ω_{ij} represents an attracting force in edge (i, j) when $\omega_{ij} > 0$ and a repelling force when $\omega_{ij} < 0$. The stress matrix $\Omega \in \mathbf{R}^{N\times N}$ is defined as

$$\Omega_{ij} = \begin{cases} 0, & \text{if } i \neq j, (i, j) \notin \mathcal{E}, \\ -\omega_{ij} & \text{if } i \neq j, (i, j) \in \mathcal{E}, \\ \sum_{k\in N_i} \omega_{ik} & \text{if } i - j, \end{cases} \tag{5.6}$$

where N_i denotes the in-neighbor set of node i. Since there are M leaders in the network, the stress matrix can be written into the following form:

$$\Omega = \begin{bmatrix} 0_{M\times M} & 0_{M\times(N-M)} \\ \Omega_{fl} & \Omega_{ff} \end{bmatrix}.$$

In this section, we need the following assumptions.

Assumption 5.1 For a nominal formation $(\mathcal{G}, r)$, $\{r_i\}_{i=1}^{i=N}$ affinely span $\mathbf{R}^d$.

This assumption means that the nominal configuration r is generic [116]. For instance, there exist at least three nodes that are not on the same line if $d = 2$ and there exist at least four nodes that are not on the same plane if $d = 3$.

Assumption 5.2 There are $d + 1$ leaders in $\mathcal{V}_l$, i.e., $M = d + 1$, and every follower in $\mathcal{V}_f$ is $(d + 1)$-reachable from $\mathcal{V}_l$.

This assumption is equivalent to that the nominal formation $(\mathcal{G}, r)$ is affinely localizable by the leaders [116].

Lemma 5.1 ([116]) *If Assumptions 5.1 and 5.2 hold, then the following statements are true.*

(i) Ω_{ff} *is positive definite.*
(ii) for any $p = [p_l^T, p_f^T]^T \in \{p \in \mathbf{R}^{dN} : p = (I_N \otimes A)r + \mathbf{1}_N \otimes b\}$, p_f *can be uniquely calculated as* $p_f = -\Omega_{ff}^{-1}\Omega_{fl}p_l$.

Let $x_l = [x_1^T, \cdots, x_M^T]^T$, $x_f = [x_{M+1}^T, \cdots, x_N^T]^T$, $x = [x_l^T, x_f^T]^T$, and $p^* = [(x_l^*)^T, (x_f^*)^T]^T$. The objective of affine formation maneuver control is to ensure that all agents converge to their desired positions in the target formation, i.e., $\lim_{t\to\infty}(x - p^*) = 0$, where p^* denotes the target position.

In light of Lemma 5.1, the desired states of followers x_f^* can be uniquely determined as $x_f^* = -(\Omega_{ff}^{-1}\Omega_{fl} \otimes I_d)x_l^*$. We can derive the tracking error of followers as

$$\delta = x_f - x_f^* = x_f + (\Omega_{ff}^{-1}\Omega_{fl} \otimes I_d)x_l^*.$$

Without loss of generality, assume that the leaders have already been steered properly, i.e., $x_l = x_l^*$. The objective of affine formation maneuver control is then converted to steer the followers such that $\delta \to 0$ as $t \to \infty$.

5.2.2 Adaptive Event-Triggered Affine Formation Controllers

Now, we are in a position to propose a fully distributed event-triggered protocol to achieve the affine formation maneuver control. Partially motivated by the results in Chap. 4, we propose the following control law for each follower:

$$\begin{aligned} u_i(t) &= \mathrm{sgn}(\sum_{j=1, j\neq i}^{N} w_{ij})(d_i + \rho_i)K\hat{\xi}_i, \\ \dot{d}_i(t) &= \hat{\xi}_i^T\Gamma\hat{\xi}_i,\ i \in \mathcal{V}_f, \end{aligned} \tag{5.7}$$

where w_{ij} denotes the (i, j)-th element of the stress matrix Ω, $\hat{\xi}_i = \sum_{k=1}^{M} w_{ik}(\hat{x}_i - \hat{x}_k) + \sum_{j=M+1}^{N} w_{ij}(\hat{x}_i - \hat{x}_j)$, $\hat{x}_j(t) = e^{A(t-t_k^j)}x_j(t_k^j)$, $\forall t \in [t_k^j, t_{k+1}^j)$, is the state estimation of agent j with t_k^j being the kth triggering instant, $\mathrm{sgn}(\cdot)$ denotes the sign function, $d_i(t)$ is the adaptive gain with initial value $d_i(0) \geq 1$, and $K = -B^TQ$, $\Gamma = QBB^TQ$ are feedback gain matrices with $Q > 0$ being the solution to the following ARE:

$$QA + A^TQ - QBB^TQ + L = 0,$$

where $L > 0$. In addition, $\rho_i = \hat{\xi}_i^TQ\hat{\xi}_i$.

The triggering condition for each agent is given as in (5.3). Once this triggering condition is satisfied, the corresponding agent broadcasts its sampled state information to its out-neighbors and updates its control law if it is a follower. Subsequently, when a follower receives updated information from its in-neighbors, it will immediately update its control law.

Theorem 5.1 *Under Assumptions 5.1 and 5.2, the proposed event-triggered control law (5.7) and the triggering function (5.3) solve the affine formation maneuver control problem for the agents in (5.1).*

Proof Denote $\xi_i = \sum_{k=1}^{M} w_{ik}(x_i - x_k) + \sum_{j=M+1}^{N} w_{ij}(x_i - x_j)$, which can be rewritten as $\xi = (\Omega_{ff} \otimes I_d)x_f + (\Omega_{fl} \otimes I_d)x_l = (\Omega_{ff} \otimes I_d)\delta$. Since Ω_{ff} is nonsingular, the objective $\delta \to 0$ can be transformed into $\xi \to 0$.

Let $\bar{d}_i$ be the virtual adaptive gain, whose dynamics are described by

$$\dot{\bar{d}}_i = \xi_i^T \Gamma \xi_i, \tag{5.8}$$

with initial value $1 \le \bar{d}_i(0) \le d_i(0)$. Define $\tilde{d}_i = d_i - \bar{d}_i$, $\bar{\rho}_i = \xi_i^T Q \xi_i$, and $\tilde{\rho}_i = \rho_i - \bar{\rho}_i$. Let $C_i = \mathrm{sgn}(\sum_{j=1, j\neq i}^{N} w_{ij})$, $C = \mathrm{diag}\{C_{M+1}, \cdots, C_N\}$, $D = \mathrm{diag}\{d_{M+1}, \cdots, d_N\}$, and $\rho = \mathrm{diag}\{\rho_{M+1}, \cdots, \rho_N\}$. The closed-loop dynamics of (5.1) in compact form can be described by

$$\begin{aligned} \dot{x}_f &= (I_{N-M} \otimes A)x_f + (I_{N-M} \otimes B)u_f \\ &= (I_{N-M} \otimes A)x_f + [C(D+\rho) \otimes BK]\hat{\xi}. \end{aligned} \tag{5.9}$$

Then, it follows that the tracking error ξ satisfies the following dynamics:

$$\begin{aligned} \dot{\xi} &= (I_{N-M} \otimes A)\xi + [\tilde{\Omega}_{ff}(D+\rho) \otimes BK]\hat{\xi} \\ &= (I_{N-M} \otimes A)\xi + [\tilde{\Omega}_{ff}(\bar{D}+\bar{\rho}) \otimes BK]\xi \\ &\quad + [\tilde{\Omega}_{ff}(\tilde{D}+\tilde{\rho}) \otimes BK]\xi + [\tilde{\Omega}_{ff}(D+\rho) \otimes BK]\tilde{\xi}, \end{aligned} \tag{5.10}$$

where $\tilde{\Omega}_{ff} = C\Omega_{ff}$ and $\bar{D}, \bar{\rho}, \tilde{D}, \tilde{\rho}$ have similar structures to D and ρ.

Construct the following Lyapunov function:

$$V_1 = \sum_{i=M+1}^{N} r_i[(\bar{d}_i + \frac{1}{2}\bar{\rho}_i)\bar{\rho}_i + \frac{(\bar{d}_i - \alpha)^2}{2} + \beta\bar{d}_i^2\bar{V}_0], \tag{5.11}$$

where $\bar{V}_0 = \sum_{j=1}^{N} \frac{\varepsilon_j}{k_j}$, r_i, α and β are positive constants. The time derivative of V_1 along the trajectories of (5.1) is given by

$$\dot{V}_1 = \sum_{i=M+1}^{N} r_i[(\bar{d}_i + \bar{\rho}_i)\dot{\bar{\rho}}_i + ((1+2\beta\bar{V}_0)\bar{d}_i - \alpha + \bar{\rho}_i)\dot{\bar{d}}_i] + \sum_{i=M+1}^{N} r_i\beta\bar{d}_i^2\dot{\bar{V}}_0$$

$$
\begin{aligned}
&= \xi^T[(\bar{D}+\bar{\rho})R \otimes (QA + A^T Q) - (\bar{D}+\bar{\rho})(R\tilde{\Omega}_{ff} + \tilde{\Omega}_{ff}^T R)(\bar{D}+\bar{\rho}) \otimes \Gamma]\xi \\
&\quad - 2\xi^T[(\bar{D}+\bar{\rho})R\tilde{\Omega}_{ff}(D+\rho) \otimes \Gamma]\tilde{\xi} - 2\xi^T[(\bar{D}+\bar{\rho})R\tilde{\Omega}_{ff}(\tilde{D}+\tilde{\rho}) \otimes \Gamma]\xi \\
&\quad + \xi^T[((2\beta\bar{V}_0+1)\bar{D} - \alpha I_{N-M} + \bar{\rho})R \otimes \Gamma]\xi \\
&\quad - \sum_{i=M+1}^{N} \beta r_i \bar{d}_i^2 \sum_{j=1}^{N}(\varepsilon_j + \frac{\sigma_j}{k_j}\tilde{x}_j^T\tilde{x}_j),
\end{aligned}
\tag{5.12}
$$

where $R = \mathrm{diag}\{r_{M+1}, \cdots, r_N\}$ is a positive-definite diagonal matrix such that $R\tilde{\Omega}_{ff} + \tilde{\Omega}_{ff}^T R > 0$.

By applying Lemma 4 in [55], we have

$$
\begin{aligned}
&- \xi^T[(\bar{D}+\bar{\rho})(R\tilde{\Omega}_{ff} + \tilde{\Omega}_{ff}^T R)(\bar{D}+\bar{\rho}) \otimes \Gamma] \\
&\quad \leq -\xi^T[\lambda_{\min}(\hat{L})(\bar{D}+\bar{\rho})^2 \otimes \Gamma]\xi,
\end{aligned}
\tag{5.13}
$$

where $\hat{L} = (R\tilde{\Omega}_{ff} + \tilde{\Omega}_{ff}^T R)$ and $\lambda_{\min}(\hat{L})$ denotes the minimum eigenvalue of $\hat{L}$.

Based on Lemma 2.5, we can obtain that

$$
\begin{aligned}
|\tilde{\rho}_i| &= \left|\hat{\xi}_i^T Q\hat{\xi}_i - \xi_i^T Q\xi_i\right| \\
&\leq \alpha_1\bar{\rho}_i + (1+\frac{1}{\alpha_1})\lambda_{\max}(Q)\tilde{\xi}_i^T\tilde{\xi}_i,
\end{aligned}
\tag{5.14}
$$

where $\alpha_1 = \frac{\lambda_{\min}(\hat{L})}{4\sigma_{\max}(R\tilde{\Omega}_{ff})}$ and $\lambda_{\max}(Q)$ denotes the largest eigenvalue of Q.

Note that

$$
\begin{aligned}
\tilde{\xi}_i^T\tilde{\xi}_i &= \sum_{j=1}^{N} w_{ij}(\tilde{x}_i - \tilde{x}_j)^T \sum_{j=1}^{N} w_{ij}(\tilde{x}_i - \tilde{x}_j) \\
&\leq 2\Omega_{ii}^2\tilde{x}_i^T\tilde{x}_i + 2\sum_{j=1}^{N} w_{ij}\tilde{x}_j^T \sum_{j=1}^{N} w_{ij}\tilde{x}_j \\
&\leq 2\Omega_{ii}^2\tilde{x}_i^T\tilde{x}_i + 2\eta_i \sum_{j=1}^{N}\tilde{x}_j^T \sum_{j=1}^{N}\tilde{x}_j \\
&\leq 2\eta_i^2\sum_{j=1}^{N}\tilde{x}_j^T\tilde{x}_j \leq 2\eta_i^2\sum_{j=1}^{N}\gamma_j\varepsilon_j(0),
\end{aligned}
\tag{5.15}
$$

where $\eta_i = \sum_{j=1, j\neq i}^{N}\left|w_{ij}\right|$ and the last inequality is obtained by using the triggering condition (5.3).

Next, we can derive from (5.14) and (5.15) that

$$|\tilde{\rho}_i| \leq \alpha_1 \bar{\rho}_i + 2(1+\frac{1}{\alpha_1})\lambda_{\max}(Q)\eta_i^2 \sum_{j=1}^{N} \gamma_j \varepsilon_j(0). \tag{5.16}$$

By using Lemma 2.5 again, we can also obtain that

$$\begin{aligned} \left|\tilde{d}_i(t)\right| &= \left|\tilde{d}_i(0) + \int_0^t [\tilde{\xi}_i^T(\tau)\Gamma\tilde{\xi}_i(\tau) + 2\tilde{\xi}_i^T(\tau)\Gamma\xi_i(\tau)]\,d\tau\right| \\ &\leq \tilde{d}_i(0) + \alpha_1\bar{d}_i - \alpha_1\bar{d}_i(0) + 2\lambda_{\max}(\Gamma)(1+\frac{1}{\alpha_1})\eta_i^2 \sum_{j=1}^{N} \frac{\gamma_j\varepsilon_j(0)}{k_j}. \end{aligned} \tag{5.17}$$

Combining (5.16) and (5.17), we have

$$\left|\tilde{d}_i + \tilde{\rho}_i\right| \leq \alpha_1(\bar{d}_i + \bar{\rho}_i) + \beta_1, \tag{5.18}$$

where

$$\beta_1 = \max_i\{\tilde{d}_i(0) + \alpha_1\bar{d}_i(0)\} + 2(1+\frac{1}{\alpha_1})\max_i\{\eta_i^2\}\sum_{j=1}^{N}(\lambda_{\max}(Q) + \frac{\lambda_{\max}(\Gamma)}{k_j})\gamma_j\varepsilon_j(0).$$

It is not difficult to obtain that

$$\begin{aligned} &-2\xi^T[(\bar{D}+\bar{\rho})R\tilde{\Omega}_{ff}(\tilde{D}+\tilde{\rho})\otimes\Gamma]\xi \\ &\leq \frac{\lambda_{\min}(\hat{L})}{4}\xi^T[(\bar{D}+\bar{\rho})^2\otimes\Gamma]\xi + \frac{\sigma_{\max}(R\tilde{\Omega}_{ff})}{\alpha_1}\xi^T[(\tilde{D}+\tilde{\rho})^2\otimes\Gamma]\xi, \end{aligned} \tag{5.19}$$

and

$$\begin{aligned} &\frac{\sigma_{\max}(R\tilde{\Omega}_{ff})}{\alpha_1}\xi^T[(\tilde{D}+\tilde{\rho})^2\otimes\Gamma]\xi \\ &\leq \frac{\sigma_{\max}(R\tilde{\Omega}_{ff})}{\alpha_1}\sum_{i=M+1}^{N}[\alpha_1(\bar{d}_i+\bar{\rho}_i)+\beta_1]^2\xi_i^T\Gamma\xi_i \\ &= \frac{\lambda_{\min}(\hat{L})}{4}\xi^T[(\bar{D}+\bar{\rho})^2\otimes\Gamma]\xi + 2\beta_1\sigma_{\max}(R\tilde{\Omega}_{ff})\xi^T[(\bar{D}+\bar{\rho})\otimes\Gamma]\xi \\ &\quad + \frac{\beta_1^2\sigma_{\max}(R\tilde{\Omega}_{ff})}{\alpha_1}\xi^T[I_{N-M}\otimes\Gamma]\xi, \end{aligned} \tag{5.20}$$

where $\sigma_{\max}(\cdot)$ represents the maximum singular value of a matrix. Substituting (5.20) into (5.19) yields

$$\begin{aligned} & -2\xi^T[(\bar{D}+\bar{\rho})R\tilde{\Omega}_{ff}(\tilde{D}+\tilde{\rho})\otimes\Gamma]\xi \\ \leq & \frac{\lambda_{\min}(\hat{L})}{2}\xi^T[(\bar{D}+\bar{\rho})^2\otimes\Gamma]\xi+2\beta_1\sigma_{\max}(R\tilde{\Omega}_{ff})\xi^T[(\bar{D}+\bar{\rho})\otimes\Gamma]\xi \\ & +\frac{\beta_1^2\sigma_{\max}(R\tilde{\Omega}_{ff})}{\alpha_1}\xi^T[I_{N-M}\otimes\Gamma]\xi. \end{aligned} \tag{5.21}$$

Similar to (5.16) and (5.17), it then follows that

$$\begin{aligned} \rho_i & \leq (1+\alpha_1)\xi_i^T Q\xi_i+(1+\frac{1}{\alpha_1})\tilde{\xi}_i^T Q\tilde{\xi}_i \\ & \leq (1+\alpha_1)\bar{\rho}_i+2\lambda_{\max}(Q)(1+\frac{1}{\alpha})\eta_i^2\sum_{j=1}^{N}\gamma_j\varepsilon_j(0), \end{aligned} \tag{5.22}$$

and

$$\begin{aligned} d_i & = d_i(0)+\int_0^t[\tilde{\xi}_i^T\Gamma\tilde{\xi}_i+2\tilde{\xi}_i^T\Gamma\xi_i^T+\xi_i^T\Gamma\xi_i]d\tau \\ & \leq d_i(0)+(1+\alpha_1)\bar{d}_i-(1+\alpha_1)\bar{d}_i(0) \\ & +2\lambda_{\max}(\Gamma)(1+\frac{1}{\alpha_1})\eta_i^2\sum_{j=1}^{N}\frac{\gamma_j\varepsilon_j(0)}{k_j}. \end{aligned} \tag{5.23}$$

Combining (5.22) and (5.23) gets the following inequality:

$$d_i+\rho_i\leq(1+\alpha_1)(\bar{d}_i+\bar{\rho}_i)+\beta_2, \tag{5.24}$$

where

$$\begin{aligned} \beta_2 = & \max_i\{d_i(0)+(1+\alpha_1)\bar{d}_i(0)\}+2(1+\frac{1}{\alpha_1})\max_i\{\eta_i^2\}\sum_{j=1}^{N}[\lambda_{\max}(Q) \\ & +\frac{\lambda_{\max}(\Gamma)}{k_j}]\gamma_j\varepsilon_j(0). \end{aligned}$$

Next, the second term in (5.12) can be scaled as follows:

$$
\begin{aligned}
& -2\xi^T[(\bar{D}+\bar{\rho})R\tilde{\Omega}_{ff}(D+\rho)\otimes\Gamma]\tilde{\xi} \\
& \le \frac{\lambda_{\min}(\hat{L})}{4}\xi^T[(\bar{D}+\bar{\rho})^2\otimes\Gamma]\xi + \frac{\sigma_{\max}(R\tilde{\Omega}_{ff})}{\alpha_1}\tilde{\xi}^T[(D+\rho)^2\otimes\Gamma]\tilde{\xi}.
\end{aligned}
\tag{5.25}
$$

Substituting (5.24) into the second term in (5.25) yields

$$
\begin{aligned}
& \frac{\sigma_{\max}(R\tilde{\Omega}_{ff})}{\alpha_1}\tilde{\xi}^T[(D+\rho)^2\otimes\Gamma]\tilde{\xi} \\
\le & \frac{\sigma_{\max}(R\tilde{\Omega}_{ff})}{\alpha_1}\sum_{i=M+1}^{N}[(1+\alpha_1)(\bar{d}_i+\bar{\rho}_i)+\beta_2]^2\tilde{\xi}_i^T\Gamma\tilde{\xi}_i \\
= & \frac{\sigma_{\max}(R\tilde{\Omega}_{ff})}{\alpha_1}\sum_{i=M+1}^{N}[(1+\alpha_1)^2\bar{d}_i^2+2(1+\alpha_1)\bar{d}_i\beta_2+\beta_2^2]\tilde{\xi}_i^T\Gamma\tilde{\xi}_i \\
& + \frac{\sigma_{\max}(R\tilde{\Omega}_{ff})}{\alpha_1}\sum_{i=M+1}^{N}[(1+\alpha_1)^2\bar{\rho}_i+2(1+\alpha_1)^2\bar{d}_i+2\beta_2(1+\alpha_1)]\bar{\rho}_i\tilde{\xi}_i^T\Gamma\tilde{\xi}_i.
\end{aligned}
\tag{5.26}
$$

Noticing that $\bar{d}_i \ge 1$, we have

$$
\begin{aligned}
& \sum_{i=M+1}^{N}[(1+\alpha_1)^2\bar{d}_i^2+2(1+\alpha_1)\bar{d}_i\beta_2+\beta_2^2]\tilde{\xi}_i^T\Gamma\tilde{\xi}_i \\
\le & \sum_{i=M+1}^{N}[(1+\alpha_1)^2+2(1+\alpha_1)\beta_2+\beta_2^2]\bar{d}_i^2\tilde{\xi}_i^T\Gamma\tilde{\xi}_i \\
\le & \, 2\sum_{i=M+1}^{N}(1+\alpha_1+\beta_2)^2\lambda_{\max}(\Gamma)\eta_i^2\bar{d}_i^2\sum_{j=1}^{N}\tilde{x}_j^T\tilde{x}_j,
\end{aligned}
\tag{5.27}
$$

and

$$
\begin{aligned}
& [(1+\alpha_1)^2\bar{\rho}_i+2(1+\alpha_1)^2\bar{d}_i+2\beta_2(1+\alpha_1)]\bar{\rho}_i\tilde{\xi}_i^T\Gamma\tilde{\xi}_i \\
\le & [(1+\alpha_1)^2\bar{\rho}_i+2(1+\alpha_1)(1+\alpha_1+\beta_2)\bar{d}_i]\bar{\rho}_i\tilde{\xi}_i^T\Gamma\tilde{\xi}_i \\
\le & \max\{(1+\alpha_1)^2, 2(1+\alpha_1)(1+\alpha_1+\beta_2)\}(\bar{d}_i+\bar{\rho}_i)\bar{\rho}_i\tilde{\xi}_i^T\Gamma\tilde{\xi}_i \\
\le & \, \alpha_2(\bar{d}_i+\bar{\rho}_i)\bar{\rho}_i\sum_{j=1}^{N}\gamma_j\varepsilon_j(t),
\end{aligned}
\tag{5.28}
$$

where $\alpha_2 = 2\max\{(1+\alpha_1)^2, 2(1+\alpha_1)(1+\alpha_1+\beta_2)\}\lambda_{\max}(\Gamma)\max_i\{\eta_i^2\}$.

Note that the triggering condition (5.3) actually indicates that $-(k_i+\sigma_i\gamma_i)\varepsilon_i \leq \dot{\varepsilon}_i \leq -k_i\varepsilon_i$. Then, it follows that $e^{-(k_i+\sigma_i\gamma_i)t}\varepsilon_i(0) \leq \varepsilon_i(t) \leq e^{-k_i t}\varepsilon_i(0)$, implying that ε_i decreases at an exponential rate. Thus, there definitely exists a time threshold T_0 such that $\alpha_2\sum_{j=1}^{N}\gamma_j\varepsilon_j(t) \leq \frac{\alpha_1\lambda_{\min}(R)\lambda_{\min}(X)}{2\sigma_{\max}[R\tilde{\Omega}_{ff}]\lambda_{\max}(Q)}$ for all $t > T_0$, where $X = -(QA + A^TQ - \Gamma) > 0$. Then, substituting (5.27) and (5.28) into (5.25), we have

$$
\begin{aligned}
& -2\xi^T[(\bar{D}+\bar{\rho})R\tilde{\Omega}_{ff}(D+\rho)\otimes\Gamma]\tilde{\xi} \\
& \leq \frac{\lambda_{\min}(\hat{L})}{4}\xi^T[(\bar{D}+\bar{\rho})^2\otimes\Gamma]\xi + \frac{1}{2}\xi^T[(\bar{D}+\bar{\rho})R\otimes X]\xi \\
& + \frac{2\sigma_{\max}(R\tilde{\Omega}_{ff})}{\alpha_1}\sum_{i=M+1}^{N}(1+\alpha_1+\beta_2)^2\lambda_{\max}(\Gamma)\eta_i^2\bar{d}_i^2\sum_{j=1}^{N}\tilde{x}_j^T\tilde{x}_j.
\end{aligned} \tag{5.29}
$$

Let $\bar{\alpha}_0 > 0$. Select $\alpha = \frac{\beta_1^2\sigma_{\max}(R\tilde{\Omega}_{ff})}{\alpha_1\lambda_{\min}(R)} + \bar{\alpha}_0$ and $\beta \geq \frac{2\sigma_{\max}(R\tilde{\Omega}_{ff})}{\min_i\{r_i\}\times\alpha_1}(1+\alpha_1+\beta_2)^2\lambda_{\max}(\Gamma)\max_i\{\eta_i^2\}\max_j\{\frac{k_j}{\sigma_j}\}$. Then, substituting (5.13), (5.21) and (5.29) into (5.12) yields

$$
\begin{aligned}
\dot{V}_1 \leq\ & \xi^T[(\bar{D}+\bar{\rho})R\otimes(QA+A^TQ)]\xi - \xi^T[\frac{\lambda_{\min}(\hat{L})}{4}(\bar{D}+\bar{\rho})^2\otimes\Gamma]\xi \\
& + 2\beta_1\sigma_{\max}(R\tilde{\Omega}_{ff})\xi^T[(\bar{D}+\bar{\rho})\otimes\Gamma]\xi + \frac{1}{2}\xi^T[(\bar{D}+\bar{\rho})R\otimes X]\xi \\
& + \xi^T[(1+2\beta\bar{V}_0(0))R(\bar{D}+\bar{\rho})\otimes\Gamma]\xi - \xi^T(\bar{\alpha}_0R\otimes\Gamma)\xi.
\end{aligned} \tag{5.30}
$$

Selecting $\bar{\beta}_0 = 1 + 2\beta\bar{V}_0(0) + \frac{2\beta_1\sigma_{\max}(R\tilde{\Omega}_{ff})}{\lambda_{\min}(R)}$, it then follows from (5.30) that

$$
\begin{aligned}
\dot{V}_1 \leq\ & \xi^T[(\bar{D}+\bar{\rho})R\otimes(QA+A^TQ+\frac{X}{2}+\bar{\beta}_0\Gamma)]\xi \\
& - \xi^T[(\frac{\lambda_{\min}(\hat{L})}{4}(\bar{D}+\bar{\rho})^2+\bar{\alpha}_0R)\otimes\Gamma]\xi.
\end{aligned}
$$

Based on Lemma 2.5 and selecting $\bar{\alpha}_0 \geq \frac{(1+\beta_0)^2\lambda_{\max}(R)}{\lambda_{\min}(\hat{L})}$, we have

$$
\begin{aligned}
\dot{V}_1 \leq\ & \xi^T[(\bar{D}+\bar{\rho})R\otimes(QA+A^TQ+\frac{X}{2}+\beta_0\Gamma)]\xi \\
& - (1+\beta_0)\xi^T[(\bar{D}+\bar{\rho})R\otimes\Gamma]\xi \\
=\ & -\frac{1}{2}\xi^T[(\bar{D}+\bar{\rho})R\otimes X]\xi \leq 0.
\end{aligned} \tag{5.31}
$$

Therefore, V_1 is bounded, which implies that ξ_i, $\bar{\rho}_i$ and $\bar{d}_i$ are bounded. Use V_1^{∞} to denote the bounded limit of V_1 when $t \to \infty$. It then follows from (5.31) that

$$\int_0^{\infty} \xi^T[(\bar{D}+\bar{\rho})R \otimes X]\xi \leq 2[V_1(0) - V_1^{\infty}].$$

It can be inferred from (5.22) and (5.23) that ρ_i and d_i are bounded, i.e., $\tilde{d}_i$ and $\tilde{\rho}_i$ are also bounded. It then can be deduced from the triggering condition (5.30) that $\tilde{x}_i$ is bounded, implying that $\tilde{\xi}_i$ is also bounded. Thus, we can deduce from (5.10) that $\dot{\xi}$ is bounded. Hence, $\xi^T[(\bar{D}+\bar{\rho})R\otimes X]\dot{\xi}$ is bounded. By using Lemma 2.4, we can conclude that $\xi^T[(\bar{D}+\bar{\rho})R \otimes X]\xi \to 0$ as $t \to \infty$, i.e., $\xi \to 0$ as $t \to \infty$. □

The following result shows that Zeno behavior does not happen.

Theorem 5.2 *Under the conditions in Theorem 5.1, the Zeno behavior is excluded.*

Proof Similar methods in Chapter 4 will be used to rule out the Zeno behavior. From the triggering condition (5.3), we have

$$\tilde{x}_i^T \tilde{x}_i \leq \gamma_i \varepsilon_i(0) e^{-k_i t}.$$

The time derivative of $\tilde{x}_i$ can be obtained as

$$\dot{\tilde{x}}_i = A\tilde{x}_i - (d_i + \rho_i)BK\hat{\xi}_i.$$

It then follows that

$$D^+\|\tilde{x}_i(t)\| \leq \|A\|\|\tilde{x}_i\| + (d_i + \rho_i)\|BK\hat{\xi}_i\|,$$

where $D^+\|\tilde{x}_i(t)\|$ is the right-hand Dini derivative of $\tilde{x}_i$. Note that d_i, ρ_i, and $\hat{\xi}_i$ are bounded, we can get that

$$\|\tilde{x}_i(t)\| \leq \frac{\Xi_i}{\|A\|}(e^{\|A\|(t-t_k^i)} - 1),$$

where Ξ_i is the upper bound of $(d_i + \rho_i)\|BK\xi_i\|$. Thus, we can infer from the triggering condition (5.3) that the triggering time t_{k+1}^i of agent i satisfies

$$\left[\frac{\Xi_i}{\|A\|}(e^{\|A\|(t_{k+1}^i - t_k^i)} - 1)\right]^2 \geq \varepsilon_i(0) e^{-(k_i + \sigma_i\gamma_i)t_{k+1}^i},$$

implying that

$$t_{k+1}^i - t_k^i \geq \frac{1}{\|A\|} \ln\left(1 + \frac{\|A\|}{\Xi_i}\sqrt{\varepsilon_i(0)e^{-(k_i+\sigma_i\gamma_i)t_{k+1}^i}}\right). \tag{5.32}$$

If Zeno behavior exists, we can find a $T > 0$ such that $t_k^i < T$ as $k \to \infty$. However, $t_{k+1}^i - t_k^i$ is strictly positive according to (5.32). Hence, $t_{k+1}^i \to \infty$ as $k \to \infty$, which means that Zeno behavior does not exist under the proposed control law and triggering condition. □

5.3 Fully Distributed Affine Formation Control in Presence of Disturbances

In practice, the agent dynamics are often subject to external disturbances such as winds. Hence, it is imperative to design event-triggered control algorithms with guaranteed robustness against external disturbances. Here, the perturbed agent dynamics are given by

$$\dot{x}_i = Ax_i + Bu_i + \varpi_i,\ i = 1, \cdots, N, \tag{5.33}$$

where $\varpi_i \in \mathbf{R}^d$ denotes the external disturbance, satisfying the following assumption:

Assumption 5.3 There exist positive constants ι and κ such that $\|\varpi_i\| \leq \iota$ for $i \in \mathcal{V}_f$ and $\|Bu_i + \varpi_i\| \leq \kappa$ for $i \in \mathcal{V}_l$.

In this case, the control law (5.7) needs to be modified to ensure that the tracking error ξ and the adaptive gain $d_i(t)$ will not diverge. Motivated by the well-known σ-modification technique [44], a fully distributed robust event-triggered affine formation controller is proposed as follows:

$$\begin{aligned} u_i(t) &= \mathrm{sgn}(\sum_{j=1, j\neq i}^{N} w_{ij})(d_i + \rho_i)K\hat{\xi}_i, \\ \dot{d}_i(t) &= \hat{\xi}_i^T \Gamma \hat{\xi}_i - \phi_i (d_i - 1)^2,\ i \in \mathcal{V}_f, \end{aligned} \tag{5.34}$$

where $\hat{\xi}_i$ and w_{ij} are defined in (5.7), ϕ_i is a small positive constant, $\rho_i = \xi_i^T P^{-1} \xi_i$, $\Gamma = P^{-1}BB^T P^{-1}$, and $P > 0$ is the solution to the following LMI:

$$AP + PA^T + \varsigma P - 2BB^T < 0, \tag{5.35}$$

where $\varsigma > 1$ is a fixed coefficient.

Theorem 5.3 *Under Assumptions 5.1–5.3, the tracking error δ of followers and the adaptive gain d_i in (5.34) are ultimately bounded under the proposed control law (5.34) and the triggering function (5.3). Moreover, the time interval between any two triggering instants is strictly positive.*

The proof of the above theorem is similar to Theorem 5.1. Interested readers can complete the proof by themselves.

5.4 Numerical Simulations and Experimental Validations

In this section, both numerical simulations and hardware experiments will be conducted to verify the effectiveness of the proposed fully distributed event-triggered formation control algorithms.

5.4.1 Numerical Simulations

We consider six agents connected by the communication topology shown in Fig. 5.1. Agents labeled as 1, 2, and 3 are leaders, and the others are followers. All the agents are assumed to have double-integrator dynamics in $\mathbf{R}^4$.

The stress matrix among the agents are constructed as follows:

$$\Omega = \begin{bmatrix} 0 & 0 & 0 & 0 & 0 & 0 \\ 0 & 0 & 0 & 0 & 0 & 0 \\ 0 & 0 & 0 & 0 & 0 & 0 \\ -\frac{2}{3} & \frac{7}{3} & -\frac{1}{3} & -\frac{4}{3} & 0 & 0 \\ -\frac{2}{3} & -\frac{1}{3} & \frac{7}{3} & 0 & -\frac{4}{3} & 0 \\ -1 & \frac{8}{5} & 0 & 0 & \frac{4}{5} & -\frac{7}{5} \end{bmatrix},$$

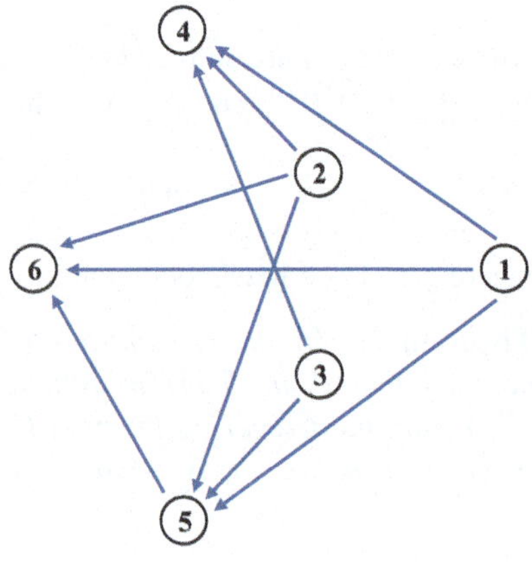

Fig. 5.1 Communication topology for simulations

while the corresponding nominal configuration of the formation is set to

$$p^T = \begin{bmatrix} 3 & 1 & 1 & 0 & 0 & -1 \\ 0 & 1 & -1 & 2 & -2 & 0 \end{bmatrix}.$$

The parameters of the controller (5.7) and the triggering function (5.3) are chosen as follows:

$$K = -\begin{bmatrix} 0.03160.075000 \\ 000.03160.0750 \end{bmatrix}, \quad \Gamma = -\begin{bmatrix} 1023.732800 \\ 23.732856.324600 \\ 001023.7328 \\ 0023.732856.3246 \end{bmatrix},$$

$$k_i = 0.08, \ \sigma_i = 0.08, \ \gamma_i = 1.$$

Triggering instants of the first 15 seconds of the formation are shown in Fig. 5.2. Clearly, the agents only need to communicate with their neighbors at discrete event instants. The consensus errors of the three followers are shown in Fig. 5.3, implying that the consensus errors will asymptotically converge to zero after every formation maneuver. Trajectories of the formation are shown in Fig. 5.4, where corresponding maneuvers can be achieved according to different scenarios.

5.4.2 Experimental Validations

In this subsubsection, we will verify the effectiveness of the proposed control schemes via some real flight experiments. The layout of our laboratory is depicted in Fig. 5.5. Our experimental equipment consists of the following three parts:

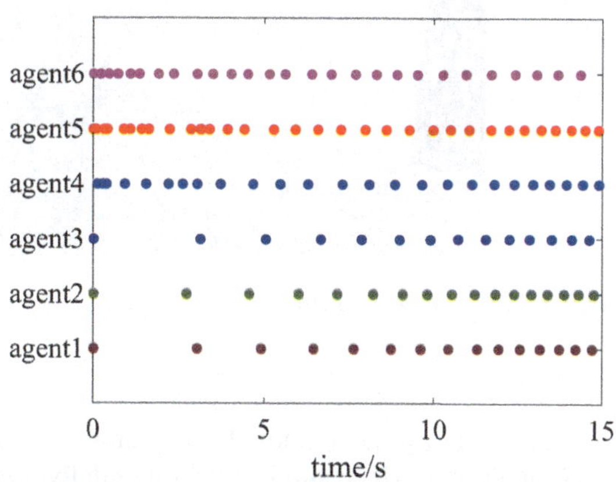

Fig. 5.2 Triggering instants of each agent

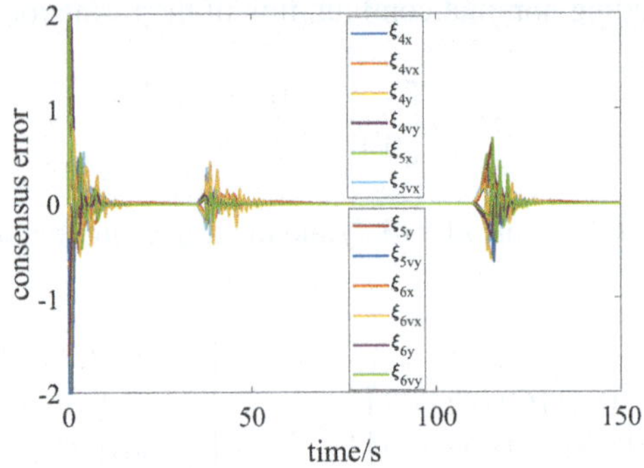

Fig. 5.3 Consensus errors of the followers

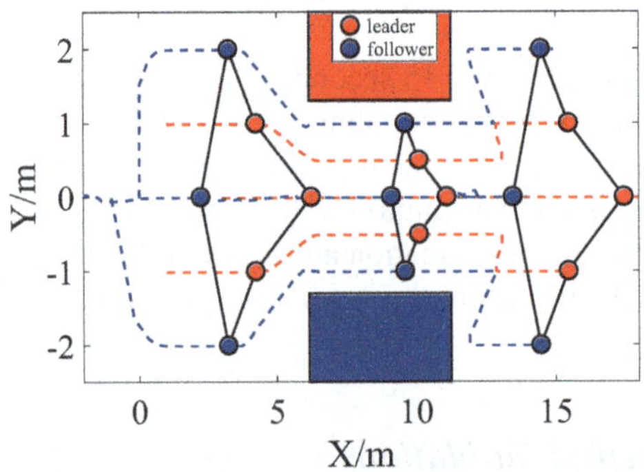

Fig. 5.4 Trajectories of the affine formation

Fig. 5.5 Layout of the laboratory

(1) The Crazyflie [5] is a nano quadrotor, which only weighs 27 g and is less than 10 cm long. It has the low-latency and long-range radio as well as Bluetooth Low Energy, which allow to steer it with Python scripts and a remote controller. Due to its small size and light weight, we can carry out a large-scale indoor experiments in a small space.

(2) The Optitrack system [79] is a set of real-time local positioning system with precision of +/−0.2mm even across large tracking areas. The Optitrack system consists of ten high-precision cameras with native frame rate of 240FPS and a Motive software which is installed on a Windows 10 personal computer to provide data integration function. Through proper initialization and calibration, we can use the Optitrack system to achieve high-accuracy real-time positioning of dynamic targets.
(3) We use Robot Operating System (ROS) to construct the experimental platform. Wi-Fi 6 communication protocol is implemented for information interactions. Python scripts are run in a ground station to steer the Crazyflies, while the Optitrack system broadcasts the Crazyflies' state information to the ground station.

The flight control experimental system is illustrated in Fig. 5.6. The Optitrack motion capture system captures the Crazyflies' position information and sends it to the host computer where Motive is installed. The Motive software integrates the data and broadcasts the information to the ground station in ROS through the Wi-Fi 6 communication protocol. Python scripts are run in the ground station utilizing position information of Crazyflies broadcast by Motive. Finally, control commands for Crazyflies are broadcast from the ground station accordingly via Crazyradio, which is a long range open USB radio dongle and capable of 2.6-GHz radio communication. In addition, there is also a remote controller for Crazyflies in case of emergencies. It is worth mentioning that the Optitrack system can only provide position information, which means that we have to exploit appropriate Kalman filtering algorithm to obtain corresponding velocity information required by our control protocols.

In this experiment, we apply a different communication topology of 7 agents shown in Fig. 5.7. Agents labeled as 1, 2, and 3 are the leaders and the rest are

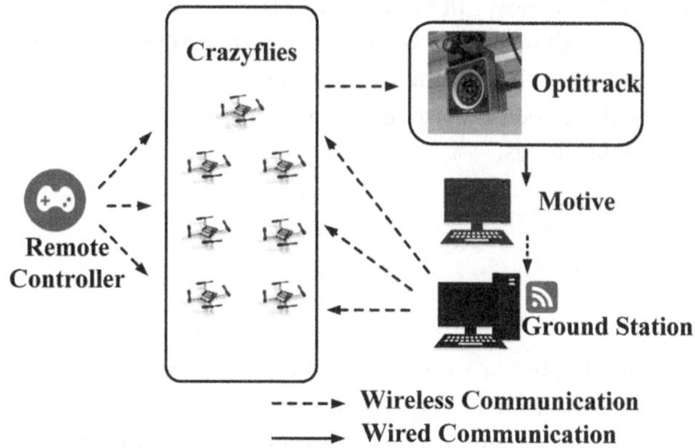

Fig. 5.6 Architecture of the flight control system

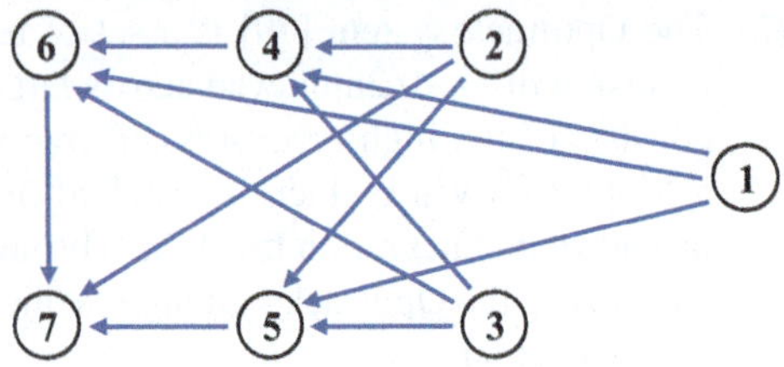

Fig. 5.7 Communication topology for experiments

followers. The corresponding stress matrix is designed as follows:

$$\Omega = \begin{bmatrix} 0 & 0 & 0 & 0 & 0 & 0 & 0 \\ 0 & 0 & 0 & 0 & 0 & 0 & 0 \\ 0 & 0 & 0 & 0 & 0 & 0 & 0 \\ -1 & 1.5 & 0.5 & -1 & 0 & 0 & 0 \\ 1 & -0.5 & -1.5 & 0 & 1 & 0 & 0 \\ 3 & -2.5 & -1.5 & -1 & 0 & 2 & 0 \\ -3 & 0.5 & 3.5 & 0 & 1 & 1 & -3 \end{bmatrix}.$$

The corresponding nominal formation configuration is set to

$$f^T = \begin{bmatrix} 2 & 1 & 1 & 0 & 0 & -1 & -1 \\ 0 & 1 & -1 & 1 & -1 & 1 & -1 \end{bmatrix}.$$

The snapshots of the actual formation at 38 s, 58 s, 78 s, and 117 s are shown in Fig. 5.8a–d, where the translational, scaling, and shearing maneuvers are depicted in Fig. 5.8b, c, and d, respectively. Figure 5.9a shows the triggering instants of seven drones in the first 23 s, which demonstrates the efficacy of the developed robust event-triggered control protocol (5.34) in reducing the number of communication interactions. Figure 5.9b shows the consensus error of the follower drones. Despite the effect of wind disturbances, the final steady-state errors of the followers still do not exceed 5 cm and the norms of the formation steady-state errors ξ_i do not exceed 0.1, demonstrating the good performance of the proposed event-triggered protocols against external disturbances.

5.5 Notes

The results of this chapter are mainly based on [8, 19]. The basic notions and definitions on affine formation are borrowed from [116]. It should be pointed out that a fully distributed event-triggered control algorithm was proposed in [78] to achieve affine formation, which however is restricted to undirected graphs. Different from the control protocol developed in [99] requiring a diagonal stabilizing gain matrix which may rely on the global information of the topology, the fully distributed

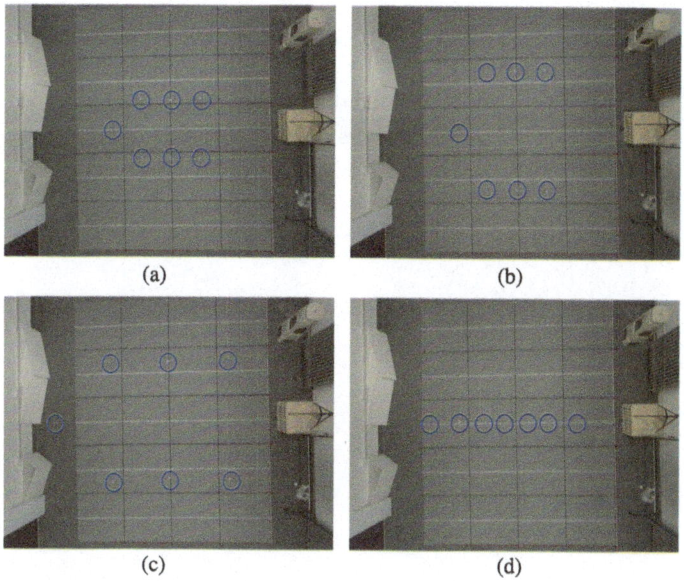

Fig. 5.8 Snapshot images of real experiments

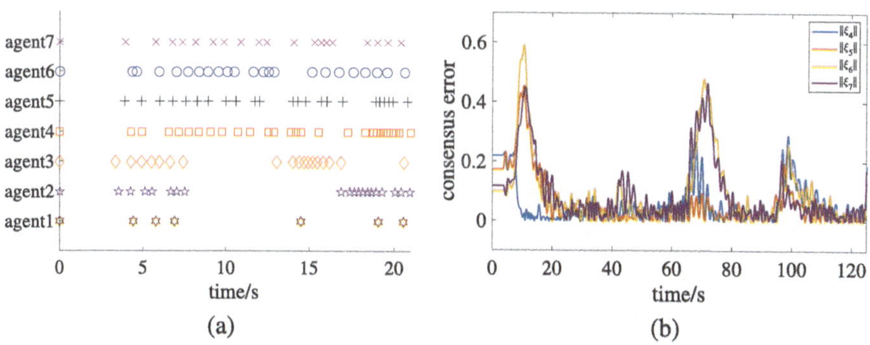

Fig. 5.9 (**a**) Triggering instants, (**b**) Consensus errors

event-triggered formation controllers proposed in this chapter depend on only local information and avoid the use of global information, by including the in-degree information of the graph.

Chapter 6
Distributed Robust Event-Triggered Control with Matching Uncertainties and Output Regulation

In Chaps. 3 and 4, we have addressed the fully distributed event-triggered consensus problems for linear multi-agent systems. Nevertheless, these proposed protocols relied on agents' accurate dynamics, which are sometimes not available as a result of the widely existing uncertainties or disturbances. This motivates us to study in this chapter distributed event-triggered control problems of multi-agent systems subject to typical time-domain uncertainties.

In Sect. 6.1, we first formulate the event-triggered consensus problem for a class of uncertain linear multi-agent systems, where the uncertainties, maybe nonlinear or time-varying, are assumed to satisfy the matching condition. Due to the existence of the nonidentical uncertainties, the multi-agent system under consideration is essentially heterogeneous. The heterogeneity nature as well as the intermittent communication render the task of designing fully distributed event-based consensus problem in this case quite challenging. Based on the relative state estimate information of neighboring agents, we design a fully distributed adaptive event-based protocol, which includes a nonlinear part to suppress the effect of uncertainties. It is shown that under the proposed protocol, consensus is achieved and the Zeno behavior is excluded.

In Sect. 6.2, we deal with the event-triggered tracking control problem for multi-agent systems with matched uncertainties in the presence of a dynamic leader whose control input is nonzero and unknown. The nonzero control input of the leader is often used to achieve certain objectives, such as to reach a desirable consensus value or to avoid hazardous obstacles. In some cases, the leader's control input might not be available to any follower, e.g., for the scenario where the leader is an uncooperative target. We establish a novel asynchronous edge-based event-triggering mechanism and propose an adaptive event-triggered protocol which includes time-varying weights into both the control laws and the triggering functions. It is shown that the followers can asymptotically track the leader as long as the leader's control input is bounded.

Z. Li et al., *Distributed Event-triggered Control*,
https://doi.org/10.1007/978-981-99-8170-0_6

In Sect. 6.3, we further devote to designing fully distributed event-triggered protocols for the cooperative output regulation problem of multi-agent networks. In this problem, there exists an exosystem, which can represent either the disturbance to be rejected or a reference signal to be tracked. The objective of cooperative output regulation is to ensure that the overall closed-loop system is asymptotically stable when the exosystem is set to zero and the tracking error approaches zero asymptotically. To handle the existence of the exosystem, we devise distributed adaptive event-triggered observers for the followers and prove that these observers asymptotically converge to the state of the exosystem. Then, we present local state feedback control inputs for the agents to regulate their outputs to zero, i.e., the cooperative output regulation problem is solved.

6.1 Robust Event-Triggered Consensus with Matching Uncertainties

6.1.1 Problem Formulation

In this section, consider a group of agents with the following dynamics:

$$\dot{x}_i = Ax_i + B[u_i + f_i(t, x_i)], \; i = 1, \cdots, N, \tag{6.1}$$

where $x_i \in \mathbf{R}^n$, $u_i \in \mathbf{R}^p$, and $f_i(t, x_i) \in \mathbf{R}^p$ represent, respectively, the state, the control input, and the uncertainty of agent i, and A, B are constant known matrices such that (A, B) is stabilizable.

Assume that there exist positive constants $\bar{f}_i$ such that $\|f_i(t, x_i)\| \leq \bar{f}_i$, $i = 1, \cdots, N$. Without loss of generality, denote $f_0 \triangleq \max\{\bar{f}_1, \cdots, \bar{f}_N\}$, which is an unknown positive constant. For simplicity, we restrict the graph $\mathcal{G}$ associated with the above N agents to be undirected in this section.

The aim of this section is to design fully distributed event-triggered protocols such that consensus can be achieved, i.e., $\lim_{t \to \infty} \|x_i(t) - x_j(t)\| = 0$, $i, j = 1, \cdots, N$, and meanwhile continuous control updating and continuous interactions can be avoided.

6.1.2 Robust Adaptive Event-Triggered Consensus Protocols

Note that we cannot use the continuous relative information between neighboring agents to design controllers due to the constraints of intermittent interactions and control updating. In this case, denote $\hat{x}_j(t) = x_j(t_k^j)$, $\forall t \in [t_k^j, t_{k+1}^j)$ as the state estimation of agent j, where t_k^j represents the k-th triggering instant of agent j.

Then, we are going to co-design the control laws and triggering functions based on $\hat{x}_j(t)$ rather than $x_j(t)$.

Let $\xi = [\xi_1^T, \cdots, \xi_N^T]^T$, where $\xi_i = \sum_{j=1}^N a_{ij}(x_i - x_j)$. Then, we have $\xi = (\mathcal{L} \otimes I_n)x$, where $x = [x_1^T, \cdots, x_N^T]^T$. Clearly, consensus is achieved if and only if ξ asymptotically converges to zero, so we can refer to ξ as the consensus error. Defining $\hat{\xi}_i(t) = \sum_{j=1}^N a_{ij}(\hat{x}_i(t) - \hat{x}_j(t))$, $i = 1, \cdots, N$, we design the following distributed adaptive event-triggered controller:

$$\begin{aligned} u_i(t) &= \hat{d}_i(t)K\hat{\xi}_i(t) + \hat{e}_i(t)g(K\hat{\xi}_i(t)), \\ \dot{d}_i(t) &= \hat{\xi}_i^T(t)\Gamma\hat{\xi}_i(t), \\ \dot{e}_i(t) &= \|K\hat{\xi}_i(t)\|, \ i = 1, \cdots, N, \end{aligned} \tag{6.2}$$

where $\hat{d}_i(t) = d_i(t_k^i)$, $\hat{e}_i(t) = e_i(t_k^i)$, $\forall t \in [t_k^i, t_{k+1}^i)$, $d_i(t)$ and $e_i(t)$ are adaptive coupling gains for agent i with initial value $d_i(0) \geq 1$ and $e_i(0) \geq 1$, respectively, K and Γ are feedback matrices to be determined, and $g(s)$ is a nonlinear function such that for $s \in \mathbf{R}^p$,

$$g(s) = \begin{cases} \frac{s}{\|s\|}, & \text{if } \|s\| \neq 0, \\ 0, & \text{if } \|s\| = 0. \end{cases} \tag{6.3}$$

The key now is to determine when to carry out communications and update controllers. Define $\tilde{x}_i(t) = \hat{x}_i(t) - x_i(t)$, $\tilde{d}_i(t) = \hat{d}_i(t) - d_i(t)$, and $\tilde{e}_i(t) = \hat{e}_i(t) - e_i(t)$ as the estimation errors. We then design a triggering condition for each agent as follows:

$$\begin{aligned} t_{k+1}^i \triangleq \inf\{t > t_k^i \mid & \tilde{x}_i^T(t)\Gamma\tilde{x}_i(t) + \sqrt{\tilde{x}_i^T(t)\Gamma\tilde{x}_i(t)} \geq \gamma_i\varepsilon_i(t) \\ & \vee \ |\tilde{d}_i(t)| \geq \theta_1 e^{-\theta_2(t-t_k^i)} \ \vee \ |\tilde{e}_i(t)| \geq \theta_3 e^{-\theta_4(t-t_k^i)}\}, \end{aligned} \tag{6.4}$$

where $t_0^i \triangleq 0$, γ_i, $i = 1, \cdots, N$, $\theta_i > 0$, $i = 1, \cdots, 4$, are positive constants, and $\varepsilon_i(t)$ is an internal variable, satisfying

$$\dot{\varepsilon}_i(t) = -k_i\varepsilon_i(t) - \sigma_i\tilde{x}_i^T(t)\Gamma\tilde{x}_i(t) - \delta_i\sqrt{\tilde{x}_i^T(t)\Gamma\tilde{x}_i(t)}, \ i = 1, \cdots, N, \tag{6.5}$$

with $\varepsilon_i(0) > 0$, $k_i > 0$, $\sigma_i > 0$, and $\delta_i > 0$. Only at event instant t_k^i, agent i sends the state value $x_i(t_k^i)$ to its neighbors. According to (6.2), agent i updates its control input $u_i(t)$ only when agent i or its neighboring agents are triggered.

Remark 6.1 It should be pointed out that in the proposed protocol (6.2)–(6.5), we introduce a linear term $\hat{d}_i(t)K\hat{\xi}_i(t)$ and a nonlinear term $\hat{e}_i(t)g(K\hat{\xi}_i(t))$, to achieve consensus and offset the effect of the matching uncertainties $f_i(t, x_i)$, respectively. To avoid relying on global information of the network, we design two adaptive gains for each agent, i.e., $d_i(t)$ and $e_i(t)$, and to avoid continuously updating the controller, we use values of these adaptive gains at event instants, i.e., $\hat{d}_i(t)K\hat{\xi}_i(t)$ and $\hat{e}_i(t)g(K\hat{\xi}_i(t))$ rather than the time-varying continuous ones. Actually, the nonlinear term can be omitted if $f_i(t, x_i) \equiv 0$, then the protocol will be reduced to the case of nominal multi-agent systems.

Letting $\tilde{\xi}_i = \sum_{j=1}^N a_{ij}(\tilde{x}_i - \tilde{x}_j)$, we obtain that $\tilde{\xi}_i = \hat{\xi}_i - \xi_i$. Denote $\bar{d}_i$ and $\bar{e}_i$ as the virtual adaptive gains, and their corresponding dynamics can be given by

$$\begin{aligned} \dot{\bar{d}}_i &= \xi_i^T \Gamma \xi_i, \\ \dot{\bar{e}}_i &= \|K\xi_i\|, \ i = 1, \cdots, N, \end{aligned} \tag{6.6}$$

with initial values $\bar{d}_i(0) \geq 1$ and $\bar{e}_i(0) \geq 1$. Further let $\check{d}_i = d_i - \bar{d}_i$ and $\check{e}_i = e_i - \bar{e}_i$. Then, we derive the closed-loop dynamics of x as follows:

$$\begin{aligned} \dot{x} = &(I_N \otimes A)x + (\bar{D} \otimes BK)\xi + (D \otimes BK)\tilde{\xi} + (\check{D} \otimes BK)\xi \\ &+ (\tilde{D} \otimes BK)\hat{\xi} + (\hat{E} \otimes B)G + (I_N \otimes B)F, \end{aligned} \tag{6.7}$$

where $\xi = [\xi_1^T, \cdots, \xi_N^T]^T$, $\tilde{\xi} = [\tilde{\xi}_1^T, \cdots, \tilde{\xi}_N^T]^T$, $\hat{\xi} = [\hat{\xi}_1^T, \cdots, \hat{\xi}_N^T]^T$, $\bar{D} = \text{diag}\{\bar{d}_1, \cdots, \bar{d}_N\}$, $D = \text{diag}\{d_1, \cdots, d_N\}$, $\tilde{D} = \text{diag}\{\tilde{d}_1, \cdots, \tilde{d}_N\}$, $\check{D} = \text{diag}\{\check{d}_1, \cdots, \check{d}_N\}$, $\hat{E} = \text{diag}\{\hat{e}_1, \cdots, \hat{e}_N\}$, $G = [g^T(K\hat{\xi}_1), \cdots, g^T(K\hat{\xi}_N)]^T$, and $F = [f_1^T, \cdots, f_N^T]^T$. Then, the closed-loop dynamics of ξ are given by

$$\begin{aligned} \dot{\xi} = &(I_N \otimes A)\xi + (\mathcal{L}\bar{D} \otimes BK)\xi + (\mathcal{L}D \otimes BK)\tilde{\xi} + (\mathcal{L}\check{D} \otimes BK)\xi \\ &+ (\mathcal{L}\tilde{D} \otimes BK)\hat{\xi} + (\mathcal{L}\hat{E} \otimes B)G + (\mathcal{L} \otimes B)F. \end{aligned} \tag{6.8}$$

The following result shows the convergence of the consensus error ξ.

Theorem 6.1 *Assuming that the graph $\mathcal{G}$ is connected, consensus can be achieved under the distributed adaptive event-triggered protocol (6.2)–(6.5) with $K = -B^TQ$ and $\Gamma = QBB^TQ$, where $P = Q^{-1}$ is a solution to the following LMI:*

$$AP + PA^T - BB^T < 0. \tag{6.9}$$

Proof Choose a Lyapunov function candidate as follows:

$$V_1 = x^T(\mathcal{L} \otimes Q)x + \sum_{i=1}^N \left[\frac{1}{2}(\bar{d}_i - \alpha_d)^2 + \beta_d \bar{d}_i V_0 + \frac{1}{2}(\bar{e}_i - \alpha_e)^2 + \beta_e \bar{e}_i V_0\right], \tag{6.10}$$

where α_d, α_e, β_d, and β_e are positive constants to be determined, and $V_0 = \sum_{j=1}^{N} \frac{\varepsilon_j}{k_j}$. By (6.4) and (6.5), we have $-(k_i + \sigma_i \gamma_i)\varepsilon_i \leq \dot{\varepsilon}_i \leq -k_i\varepsilon_i$, and thus $e^{-(k_i+\sigma_i\gamma_i)t}\varepsilon_i(0) \leq \varepsilon_i(t) \leq e^{-k_i t}\varepsilon_i(0)$. The time derivative of V_1 along the trajectory of (6.7) can be written as

$$\begin{aligned}\dot{V}_1 &= x^T[\mathcal{L} \otimes (QA + A^T Q)]x - 2\xi^T(\bar{D} \otimes \Gamma)\xi - 2\xi^T(D \otimes \Gamma)\tilde{\xi} \\ &\quad - 2\xi^T(\check{D} \otimes \Gamma)\xi - 2\xi^T(\tilde{D} \otimes \Gamma)\hat{\xi} + 2\xi^T(\hat{E} \otimes QB)G + 2\xi^T(I_N \otimes QB)F \\ &\quad + \sum_{i=1}^{N}(\bar{d}_i - \alpha_d + \beta_d V_0)\xi_i^T\Gamma\xi_i + \sum_{i=1}^{N}(\bar{e}_i - \alpha_e + \beta_e V_0)\|K\xi_i\| \\ &\quad - \sum_{i=1}^{N}(\beta_d\bar{d}_i + \beta_e\bar{e}_i)\sum_{j=1}^{N}\left(\varepsilon_j + \frac{\sigma_j}{k_j}\tilde{x}_j^T\Gamma\tilde{x}_j + \frac{\delta_j}{k_j}\sqrt{\tilde{x}_j^T\Gamma\tilde{x}_j}\right).\end{aligned} \tag{6.11}$$

Next, we focus on analyzing (6.11).

(1) We first deal with the term $-2\xi^T(D \otimes \Gamma)\tilde{\xi}$. By Lemma 2.5, we obtain that

$$\begin{aligned}-2\xi^T(D \otimes \Gamma)\tilde{\xi} &\leq \frac{1}{2}\xi^T(D \otimes \Gamma)\xi + 2\tilde{\xi}^T(D \otimes \Gamma)\tilde{\xi} \\ &= \frac{1}{2}\xi^T(\bar{D} \otimes \Gamma)\xi + \frac{1}{2}\xi^T(\check{D} \otimes \Gamma)\xi \\ &\quad + 2\tilde{\xi}^T(\bar{D} \otimes \Gamma)\tilde{\xi} + 2\tilde{\xi}^T(\check{D} \otimes \Gamma)\tilde{\xi}.\end{aligned}$$

Note that

$$\begin{aligned}\check{d}_i(t) &= \int_0^t (\hat{\xi}_i^T(\tau)\Gamma\hat{\xi}_i(\tau) - \xi_i^T(\tau)\Gamma\xi_i(\tau))d\tau + \check{d}_i(0) \\ &= \int_0^t (\tilde{\xi}_i^T(\tau)\Gamma\tilde{\xi}_i(\tau) + 2\xi_i^T(\tau)\Gamma\tilde{\xi}_i(\tau))d\tau + \check{d}_i(0) \\ &\leq \int_0^t \xi_i^T(\tau)\Gamma\xi_i(\tau)d\tau + 2\int_0^t \tilde{\xi}_i^T(\tau)\Gamma\tilde{\xi}_i(\tau)d\tau + \check{d}_i(0) \\ &= \bar{d}_i(t) + 2\int_0^t \tilde{\xi}_i^T(\tau)\Gamma\tilde{\xi}_i(\tau)d\tau + \check{d}_i(0) - \bar{d}_i(0),\end{aligned} \tag{6.12}$$

and

$$\begin{aligned}\tilde{\xi}_i^T\Gamma\tilde{\xi}_i &= \sum_{j=1}^{N} a_{ij}(\tilde{x}_i-\tilde{x}_j)^T\Gamma\sum_{j=1}^{N} a_{ij}(\tilde{x}_i-\tilde{x}_j)\\ &\leq 2l_{ii}^2\tilde{x}_i^T\Gamma\tilde{x}_i + 2\sum_{j=1}^{N} a_{ij}\tilde{x}_j^T\Gamma\sum_{j=1}^{N} a_{ij}\tilde{x}_j\\ &\leq 2l_{ii}^2\sum_{j=1}^{N}\tilde{x}_j^T\Gamma\tilde{x}_j.\end{aligned} \tag{6.13}$$

Combining (6.12) and (6.13), we get that

$$\begin{aligned}\check{d}_i(t) &\leq \bar{d}_i(t)-\bar{d}_i(0)+\check{d}_i(0)+4l_{ii}^2\sum_{j=1}^{N}\int_0^t \tilde{x}_j^T(\tau)\Gamma\tilde{x}_j(\tau)d\tau\\ &\leq \bar{d}_i(t)-\bar{d}_i(0)+\check{d}_i(0)+4l_{ii}^2\sum_{j=1}^{N}\int_0^\infty \gamma_j\varepsilon_j(\tau)d\tau\\ &\leq \bar{d}_i(t)-\bar{d}_i(0)+\check{d}_i(0)+4l_{ii}^2\sum_{j=1}^{N}\frac{\gamma_j\varepsilon_j(0)}{k_j}.\end{aligned} \tag{6.14}$$

Thus, we have

$$2\tilde{\xi}^T(\check{D}\otimes\Gamma)\tilde{\xi} = \sum_{i=1}^{N} 2\check{d}_i\tilde{\xi}_i^T\Gamma\tilde{\xi}_i \leq \sum_{i=1}^{N}\left(2\bar{d}_i+\beta_1\right)\tilde{\xi}_i^T\Gamma\tilde{\xi}_i,$$

where $\beta_1 \geq \max\{-2\bar{d}_i(0)+2\check{d}_i(0)+8l_{ii}^2\sum_{j=1}^{N}\frac{\gamma_j\varepsilon_j(0)}{k_j}\}$. Then, it is easy to derive that

$$\begin{aligned}-2\xi^T(D\otimes\Gamma)\tilde{\xi} &\leq \frac{1}{2}\xi^T(\bar{D}\otimes\Gamma)\xi + \frac{1}{2}\xi^T(\check{D}\otimes\Gamma)\xi\\ &\quad+\sum_{i=1}^{N}\left(8\bar{d}_i+2\beta_1\right)l_{ii}^2\sum_{j=1}^{N}\tilde{x}_j^T\Gamma\tilde{x}_j.\end{aligned} \tag{6.15}$$

(2) Consider the term $-\frac{3}{2}\xi^T(\check{D}\otimes\Gamma)\xi$. Obviously, we have

$$\begin{aligned}
-\check{d}_i(t) &= -\check{d}_i(0) + \int_0^t (\xi_i^T(\tau)\Gamma\xi_i(\tau) - \hat{\xi}_i^T(\tau)\Gamma\hat{\xi}_i(\tau))d\tau \\
&= -\check{d}_i(0) + \int_0^t (-\tilde{\xi}_i^T(\tau)\Gamma\tilde{\xi}_i(\tau) - 2\xi_i^T(\tau)\Gamma\tilde{\xi}_i(\tau))d\tau \\
&\le -\check{d}_i(0) + \frac{1}{3}\int_0^t \xi_i^T(\tau)\Gamma\xi_i(\tau)d\tau + 2\int_0^t \tilde{\xi}_i^T(\tau)\Gamma\tilde{\xi}_i(\tau)d\tau \\
&\le -\check{d}_i(0) + \frac{1}{3}\bar{d}_i(t) - \frac{1}{3}\bar{d}_i(0) + 4l_{ii}^2\sum_{j=1}^N \frac{\gamma_j\varepsilon_j(0)}{k_j}.
\end{aligned}$$

Thus, we can obtain that

$$\begin{aligned}
-\frac{3}{2}\xi^T(\check{D}\otimes\Gamma)\xi &= \sum_{i=1}^N -\frac{3}{2}\check{d}_i\xi_i^T\Gamma\xi_i \\
\le \sum_{i=1}^N \Big(\frac{1}{2}\bar{d}_i(t) &- \frac{3}{2}(\check{d}_i(0) + \frac{1}{3}\bar{d}_i(0)) + 6\sum_{i=1}^N l_{ii}^2\sum_{j=1}^N \frac{\gamma_i\varepsilon_j(0)}{k_j}\Big)\xi_i^T\Gamma\xi_i \\
\le \frac{1}{2}\xi^T(\bar{D}\otimes\Gamma)\xi &+ \xi^T(\beta_2 I_N\otimes\Gamma)\xi,
\end{aligned} \tag{6.16}$$

where $\beta_2 \ge \max\{-\frac{3}{2}(\check{d}_i(0) + \frac{1}{3}\bar{d}_i(0)) + 6\sum_{i=1}^N l_{ii}^2\sum_{j=1}^N \frac{\gamma_i\varepsilon_j(0)}{k_j}\}$.

(3) Consider the term $-2\xi^T(\tilde{D}\otimes\Gamma)\hat{\xi}$. According to the triggering rule (6.4), we have

$$\begin{aligned}
-2\xi^T(\tilde{D}\otimes\Gamma)\hat{\xi} &= -2\sum_{i=1}^N \tilde{d}_i\xi^T\Gamma\hat{\xi}_i \\
&\le \theta_1\sum_{i=1}^N \xi_i^T\Gamma\xi_i + \theta_1\sum_{i=1}^N \hat{\xi}_i^T\Gamma\hat{\xi}_i \\
&\le 3\theta_1\sum_{i=1}^N \xi_i^T\Gamma\xi_i + 2\theta_1\sum_{i=1}^N \tilde{\xi}_i^T\Gamma\tilde{\xi}_i \\
&\le 3\theta_1\sum_{i=1}^N \xi_i^T\Gamma\xi_i + 4\theta_1\sum_{i=1}^N l_{ii}^2\sum_{j=1}^N \tilde{x}_j^T\Gamma\tilde{x}_j.
\end{aligned} \tag{6.17}$$

(4) It is not difficult to derive that

$$
\begin{aligned}
2\xi^T(\hat{E}\otimes QB)G &= 2\sum_{i=1}^{N}\hat{e}_i\xi_i^T QBg(K\hat{\xi}_i)\\
&= 2\sum_{i=1}^{N}\hat{e}_i\xi_i^T QB\frac{-B^TQ\hat{\xi}_i}{\|B^TQ\hat{\xi}_i\|}\\
&= -2\sum_{i=1}^{N}\hat{e}_i\|B^TQ\hat{\xi}_i\| + 2\sum_{i=1}^{N}\hat{e}_i\tilde{\xi}_i^T QB\frac{B^TQ\hat{\xi}_i}{\|B^TQ\hat{\xi}_i\|}\\
&\le -2\sum_{i=1}^{N}\hat{e}_i\|B^TQ\hat{\xi}_i\| + 2\sum_{i=1}^{N}\hat{e}_i\|B^TQ\tilde{\xi}_i\|\\
&\le -2\sum_{i=1}^{N}\hat{e}_i\|B^TQ\xi_i\| + 4\sum_{i=1}^{N}\hat{e}_i\|B^TQ\tilde{\xi}_i\|.
\end{aligned}
\tag{6.18}
$$

Note that

$$
\begin{aligned}
-\check{e}_i(t) &= -\int_0^t(\|K\hat{\xi}_i(\tau)\| - \|K\xi_i(\tau)\|)d\tau - \check{e}_i(0)\\
&\le \int_0^t(\frac{1}{2}\|K\xi_i(\tau)\| - \frac{1}{2}\|K\hat{\xi}_i(\tau)\| + \frac{1}{2}\|K\tilde{\xi}_i(\tau)\|)d\tau - \check{e}_i(0)\\
&\le \frac{1}{2}\bar{e}_i(t) - \frac{1}{2}\bar{e}_i(0) - \check{e}_i(0) + \frac{1}{2}\int_0^t\|K\tilde{\xi}_i(\tau)\|d(\tau),
\end{aligned}
\tag{6.19}
$$

where

$$
\begin{aligned}
\frac{1}{2}\int_0^t\|K\tilde{\xi}_i(\tau)\|d(\tau) &\le \frac{1}{2}\int_0^t\sqrt{2l_{ii}^2\sum_{j=1}^{N}\tilde{x}_j^T(\tau)\Gamma\tilde{x}_j(\tau)d(\tau)}\\
&\le \frac{\sqrt{2}}{2}l_{ii}\int_0^t\sqrt{\sum_{j=1}^{N}\gamma_j\varepsilon_j(\tau)d(\tau)}\\
&\le \frac{\sqrt{2}}{2}l_{ii}\sum_{j=1}^{N}\sqrt{\int_0^t\gamma_j\varepsilon_j(\tau)d(\tau)}\\
&\le \frac{\sqrt{2}}{2}l_{ii}\sum_{j=1}^{N}\sqrt{\frac{\gamma_j\varepsilon_j(0)}{k_j}}.
\end{aligned}
\tag{6.20}
$$

Substituting (6.20) into (6.19) yields

$$-\check{e}_i(t) \le \frac{1}{2}\bar{e}_i(t) - \frac{1}{2}\bar{e}_i(0) - \check{e}_i(0) + \frac{\sqrt{2}}{2}l_{ii}\sum_{j=1}^{N}\sqrt{\frac{\gamma_j\varepsilon_j(0)}{k_j}}. \tag{6.21}$$

According to (6.19), we then have

$$\begin{aligned} -2\sum_{i=1}^{N}\hat{e}_i\|B^TQ\xi_i\| &= -2\sum_{i=1}^{N}(\bar{e}_i+\tilde{e}_i+\check{e}_i)\|B^TQ\xi_i\| \\ &\le -\sum_{i=1}^{N}\bar{e}_i(t)\|B^TQ\xi_i\| + \sum_{i=1}^{N}\beta_3\|B^TQ\xi_i\|, \end{aligned} \tag{6.22}$$

where $\beta_3 \ge \max\{2\theta_3 - \bar{e}_i(0) - 2\check{e}_i(0) + \sqrt{2}l_{ii}\sum_{j=1}^{N}\sqrt{\frac{\gamma_j\varepsilon_j(0)}{k_j}}\}$.

It is not difficult to see that

$$\begin{aligned} \check{e}_i(t) &= \int_0^t (\|K\hat{\xi}_i(\tau)\| - \|K\xi_i(\tau)\|)d\tau + \check{e}_i(0) \\ &\le \int_0^t \|K\tilde{\xi}_i(\tau)\|d\tau + \check{e}_i(0) \\ &\le \sqrt{2}l_{ii}\sum_{j=1}^{N}\sqrt{\frac{\gamma_j\varepsilon_j(0)}{k_j}} + \check{e}_i(0). \end{aligned} \tag{6.23}$$

According to (6.23), we have

$$\begin{aligned} &4\sum_{i=1}^{N}\hat{e}_i\|B^TQ\tilde{\xi}_i\| \le 4\sum_{i=1}^{N}(\bar{e}_i+\tilde{e}_i+\check{e}_i)\|B^TQ\tilde{\xi}_i\| \\ &\le 4(1+\theta_3+\check{e}_i(0)+\sqrt{2}l_{ii}\sum_{j=1}^{N}\sqrt{\frac{\gamma_j\varepsilon_j(0)}{k_j}})\sum_{i=1}^{N}\bar{e}_i\|B^TQ\tilde{\xi}_i\| \\ &\le \beta_4\sum_{i=1}^{N}\bar{e}_i l_{ii}\sum_{j=1}^{N}\sqrt{\tilde{x}_j\Gamma\tilde{x}_j}, \end{aligned} \tag{6.24}$$

where $\beta_4 \ge \max\{4(1+\theta_3+\check{e}_i(0)+\sqrt{2}l_{ii}\sum_{j=1}^{N}\sqrt{\frac{\gamma_j\varepsilon_j(0)}{k_j}})\}$.

Substituting (6.22) and (6.24) into (6.18) yields

$$\begin{aligned} 2\xi^T(\hat{E} \otimes QB)G \leq & -\sum_{i=1}^{N} \bar{e}_i(t)\|B^T Q\xi_i\| + \sum_{i=1}^{N} \beta_3 \|B^T Q\xi_i\| \\ & + \sum_{i=1}^{N} \beta_4 \bar{e}_i l_{ii} \sum_{j=1}^{N} \sqrt{\tilde{x}_j \Gamma \tilde{x}_j}. \end{aligned} \tag{6.25}$$

(5) We can also obtain that

$$\begin{aligned} 2\xi^T(I_N \otimes QB)F & \leq 2\sum_{i=1}^{N} \|B^T Q\xi_i\| \cdot \|f_i\| \\ & \leq 2f_0 \sum_{i=1}^{N} \|B^T Q\xi_i\|. \end{aligned} \tag{6.26}$$

According to the above analyses, substituting (6.15), (6.16), (6.17), (6.25), and (6.26) into (6.11) yields

$$\begin{aligned} \dot{V}_1 \leq & \; x^T[\mathcal{L} \otimes (QA + A^T Q)]x + \xi^T[(\beta_2 + 3\theta_1)I_N \otimes \Gamma]\xi \\ & + \sum_{i=1}^{N}((-\alpha_d + \beta_d V_0)\xi_i^T \Gamma \xi_i + (-\alpha_e + \beta_e V_0)\|K\xi_i\|) \\ & + (\beta_3 + 2f_0)\sum_{i=1}^{N} \|B^T Q\xi_i\| + \beta_4 \sum_{i=1}^{N} \bar{e}_i l_{ii} \sum_{j=1}^{N} \sqrt{\tilde{x}_j \Gamma \tilde{x}_j} \\ & + \sum_{i=1}^{N}(8\bar{d}_i + 2\beta_1 + 4\theta_1) l_{ii}^2 \sum_{j=1}^{N} \tilde{x}_j^T \Gamma \tilde{x}_j \\ & - \sum_{i=1}^{N}(\beta_d \bar{d}_i + \beta_e \bar{e}_i)\sum_{j=1}^{N}(\varepsilon_j + \frac{\sigma_j}{k_j}\tilde{x}_j^T \Gamma \tilde{x}_j + \frac{\delta_j}{k_j}\sqrt{\tilde{x}_j^T \Gamma \tilde{x}_j}). \end{aligned} \tag{6.27}$$

By choosing $\beta_d \geq 2(4+\beta_1+2\theta_1)\max\{l_{ii}^2\}\max\{\frac{k_i}{\sigma_i}\}$, $\beta_e \geq \beta_4 \max\{l_{ii}\}\max\{\frac{k_i}{\delta_i}\}$, $\alpha_d = \alpha_1 + \beta_d V_0(0) + \beta_2 + 3\theta_1$, and $\alpha_e = \beta_e V_0(0) + \beta_3 + 2f_0$, we get from (6.27) that

$$
\begin{aligned}
\dot{V}_1 &\leq x^T[\mathcal{L}\otimes(QA + A^TQ)]x - \xi^T(\alpha_1 I_N \otimes \Gamma)\xi \\
&\leq x^T[\mathcal{L}\otimes(QA + A^TQ - \Gamma)]x \\
&\leq \frac{1}{\Gamma_N(\mathcal{L})}\xi^T[I_N \otimes (QA + A^TQ - \Gamma)]\xi \\
&\leq 0,
\end{aligned}
\tag{6.28}
$$

where the second inequality is obtained by applying Lemma 2.2 and choosing $\alpha_1 \geq \frac{1}{\Gamma_2(\mathcal{L})}$, $\Gamma_2(\mathcal{L})$ and $\Gamma_N(\mathcal{L})$ denote the smallest and largest nonzero eigenvalues of $\mathcal{L}$, respectively.

Therefore, $V_1(t)$ is bounded, and so are ξ_i, $\bar{d}_i$, and $\bar{e}_i$. Moreover, V_1 has a finite limit as $t \to \infty$, denoted by V_1^∞. According to (6.28), we can further obtain that

$$
\int_0^\infty \xi^T[I_N \otimes (-(QA + A^TQ - \Gamma))]\xi dt \leq \Gamma_N(\mathcal{L})(V_1(0) - V_1^\infty).
$$

On the other hand, by (6.14) and (6.23), we can get that $d_i = \check{d}_i + \bar{d}_i \leq 2\bar{d}_i - \bar{d}_i(0) + d_i(0) + 4l_{ii}^2 \sum_{j=1}^N \frac{\gamma_j \varepsilon_j(0)}{k_j}$ and $e_i = \check{e}_i + \bar{e}_i \leq \bar{e}_i + \check{e}_i(0) + \sqrt{2}l_{ii}\sum_{j=1}^N \sqrt{\frac{\gamma_j\varepsilon_j(0)}{k_j}}$, implying that d_i and e_i are bounded. Then, $\check{d}_i = d_i - \bar{d}_i$ is also bounded. According to the triggering rule, $\tilde{x}_i^T\Gamma\tilde{x}_i$ is bounded, implying that $(I_N \otimes BK)\tilde{x}$ is bounded. Following this, we obtain that $(I_N \otimes BK)\tilde{\xi}$ is bounded, which further implies that $(I_N \otimes BK)\hat{\xi} = (I_N \otimes BK)(\xi + \tilde{\xi})$ is bounded. Noting that $\hat{e}_i = e_i + \tilde{e}_i$ is bounded, we can obtain that $(\mathcal{L}\hat{E} \otimes B)G$ is bounded. Also noting that $(\mathcal{L} \otimes B)F$ is bounded, according to (6.7), we have

$$
\begin{aligned}
\dot{\xi} &= (I_N \otimes A)\xi + (\mathcal{L}\bar{D} \otimes BK)\xi + (\mathcal{L}D \otimes BK)\tilde{\xi} + (\mathcal{L}\check{D} \otimes BK)\xi \\
&\quad + (\mathcal{L}\tilde{D} \otimes BK)\hat{\xi} + (\mathcal{L}\hat{E} \otimes B)G + (\mathcal{L} \otimes B)F,
\end{aligned}
\tag{6.29}
$$

which is bounded. It follows from Lemma 2.4 that $\xi^T[I_N \otimes (-(QA + A^TQ - \Gamma))]\xi \to 0$ as $t \to \infty$. Thus, $\xi \to 0$ as $t \to \infty$, i.e., consensus is achieved. □

Remark 6.2 It should be mentioned that this section is substantially different from [13], in which event-triggered protocols were also designed to achieve consensus of multi-agent systems with the existence of matched uncertainties and limited frequency of communications. On one hand, the task here is to design a fully distributed protocol to avoid both continuous interaction among neighbors and continuous updating of controllers, which is more complex. On the other hand, the methods used are essentially different. Particularly, to determine when to carry out communications, the work [13] designed static triggering conditions for agents, which include two state-based thresholds depending on the local information and a time-based threshold. However, in this section, we design a novel dynamic event-triggering condition without such state-based thresholds, as a result of which the triggering condition described by (6.4)–(6.5) is decoupled from the network topology. Obviously, it is easier for each agent to compute the triggering condition (6.4)–(6.5) and determine event instants compared with [13].

Remark 6.3 Different from [106] in which the authors proposed dynamic event-triggered protocols for multiple single integrators, the dynamic adaptive event-based algorithm given in this section is applicable to general linear networks. In [58], the authors designed distributed adaptive event-based algorithms for general linear multi-agent systems in which the controller $u_i(t)$ relies on the time-varying weight $c_i(t)$ and the triggering condition requires continuous relative information. Different from the protocols proposed in [58], the protocol here requires neither continuous interactions among neighboring agents nor continuous updating of controllers. Different from the adaptive event-triggered protocol in [81] designed for nominal systems, the protocol in this section is applicable to the multi-agent systems with matching uncertainties. It should be admitted that this work is partly inspired by Chap. 4, in which the adaptive dynamic event-based protocols were designed for the multi-agent systems with general directed graphs. The main difference is that the proposed protocols in this section are applicable to uncertain multi-agent systems.

Theorem 6.2 *Under the conditions in Theorem 6.1, Zeno behavior is excluded.*

Proof We rule out the Zeno behavior by providing a contradiction argument. Without loss of generality, suppose that agent i exists Zeno behavior, i.e., there exists $T < +\infty$ such that $\lim_{k\to\infty} t_k^i = T$. Then, for any positive constant Δ, there exists a positive integer k_0 such that for $k \geq k_0$, $t_k^i \in [T-\Delta, T)$.

In light of Theorem 6.1, we obtain that x_i, $\hat{\xi}_i$, $\hat{d}_i$, and $\hat{e}_i$ are all bounded for time during $[0, T)$. Noting that $\tilde{x}_i(t_{k0}^i) = 0$, $\tilde{d}_i(t_{k0}^i) = 0$, and $\tilde{e}_i(t_{k0}^i) = 0$, we have

$$
\begin{aligned}
\tilde{x}_i^T(t)\Gamma\tilde{x}_i(t) &= \left[\int_{t_{k0}^i}^{t}\Big(Ax_i(\tau) + Bf_i(\tau, x_i(\tau)) + BK\hat{d}_i(\tau)\hat{\xi}_i(\tau)\Big)\,d\tau\right]^T \Gamma \\
&\quad \times \left[\int_{t_{k0}^i}^{t}\Big(Ax_i(\tau) + Bf_i(\tau, x_i(\tau)) + BK\hat{d}_i(\tau)\hat{\xi}_i(\tau)\Big)\,d\tau\right] \\
&\leq \|\Gamma\|\rho_1^2(t - t_{k0}^i)^2, \\
|\tilde{d}_i(t)| &= \left|\int_{t_{k0}^i}^{t} \hat{\xi}_i^T(\tau)\Gamma\hat{\xi}_i(\tau)d\tau\right| \leq \rho_2(t - t_{k0}^i), \\
|\tilde{e}_i(t)| &= \left|\int_{t_{k0}^i}^{t} \|K\hat{\xi}_i(\tau)\|d\tau\right| \leq \rho_3(t - t_{k0}^i),
\end{aligned}
\tag{6.30}
$$

where ρ_1, ρ_2, and ρ_3 denote the upper bounds of $\|Ax_i(t) + Bf_i(t, x_i(t)) + BK\hat{d}_i(t)\hat{\xi}_i(t)\|$, $\hat{\xi}_i^T(t)\Gamma\hat{\xi}_i(t)$, and $\|K\hat{\xi}_i(t)\|$ for $t \in [0, T]$, respectively.

Denote $\Delta \triangleq \min\{\Delta_1, \Delta_2, \Delta_3\}$, where $\Delta_1 = \sqrt{\frac{\gamma_i e^{-(k_i+\sigma_i\gamma_i)T}}{\|\Gamma\|\rho_1^2}}$, $\Delta_2 = \frac{\theta_1 e^{-\theta_2 T}}{\rho_2}$, and $\Delta_3 = \frac{\theta_3 e^{-\theta_4 T}}{\rho_3}$. Under the triggering rule (6.4), the next event for agent i after $t_{k_0}^i$ will happen first when

$$
\begin{aligned}
&\tilde{x}_i(t)\Gamma\tilde{x}_i(t) \geq \gamma_i\varepsilon_i(t) \geq \gamma_i e^{-(k_i+\sigma_i\gamma_i)T} \\
&\vee\ |\tilde{d}_i(t)| \geq \theta_1 e^{-\theta_2(t-t_k^i)} \geq \theta_1 e^{-\theta_2 T} \\
&\vee\ |\tilde{e}_i(t)| \geq \theta_3 e^{-\theta_4(t-t_k^i)} \geq \theta_3 e^{-\theta_4 T}.
\end{aligned}
\tag{6.31}
$$

It follows that $t_{k_0+1}^i - t_{k_0}^i \geq \Delta$. Thus, we have $t_{k_0+1}^i \geq T$, which does not consist with $t_{k_0+1}^i < T$. Therefore, there does not exist the Zeno behavior. □

Remark 6.4 We have shown that the proposed event-triggered protocol can guarantee the asymptotic convergence of the consensus error for multi-agent systems in the presence of matching uncertainties. In some practical occasions, there may exist non-matching uncertainties. For example, consider the following system dynamics:

$$
\dot{x}_i = Ax_i + B[u_i + f_i(t, x_i)] + w_i,\ i = 1, \cdots, N, \tag{6.32}
$$

where $w_i \in \mathbf{R}^n$ are non-matching uncertainties satisfying $\|w_i\| \leq v_i$ with v_i, $i = 1, \cdots, N$, being positive constants. In this case, we need to use the σ-modification technique [44, 53] for the protocol (6.2) as follows:

$$
\begin{aligned}
u_i(t) &= \hat{d}_i(t)K\hat{\xi}_i(t) + \hat{e}_i(t)g(K\hat{\xi}_i(t)), \\
\dot{d}_i(t) &= -\varphi\hat{d}_i(t) + \hat{\xi}_i^T(t)\Gamma\hat{\xi}_i(t), \\
\dot{e}_i(t) &= -\mathcal{A}\hat{e}_i(t) + \|K\hat{\xi}_i(t)\|,\ i = 1, \cdots, N,
\end{aligned}
\tag{6.33}
$$

where φ and $\mathcal{A}$ are small positive constants. The events are still determined by the triggering condition (6.4). The consensus error ξ can be shown to converge to a bounded residual set [53, 65]. The details are omitted for conciseness.

6.1.3 Simulation Results

In this subsection, a numerical example will be presented to verify the effectiveness of the established theoretical results.

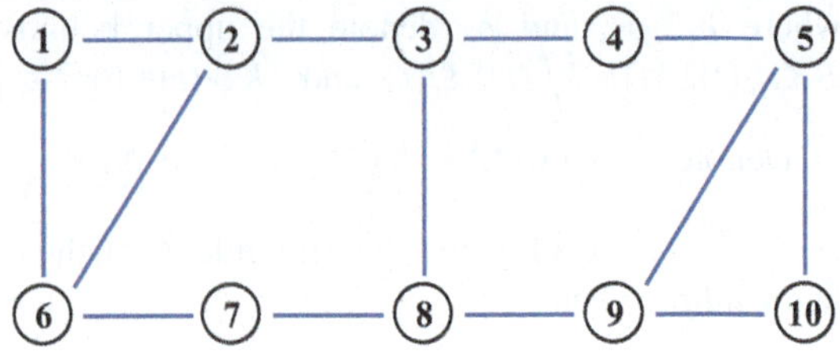

Fig. 6.1 An undirected communication graph

Consider a multi-agent system with ten agents satisfying (6.1), where

$$A = \begin{bmatrix} 0 & 1 \\ 0 & 0 \end{bmatrix}, \quad B = \begin{bmatrix} 0 \\ 1 \end{bmatrix},$$

$$f_i(t, x_i) = i/10 \sin(t), \ i = 1, 2, 3, \ f_i(t, x_i) = ie^{-t}, \ i = 4, 5, 6,$$

$$f_i(t, x_i) = i/(i + t), \ i = 7, 8, \ f_i(t, x_i) = \sin(\|x_i(t)\|)/i, \ i = 9, 10.$$

The associated topology is given in Fig. 6.1.

We use the dynamic event-triggered protocol composed of (6.2)–(6.5) for the above multi-agent systems with matching uncertainties. The parameters in the protocol are chosen as $\theta_i = 1, i = 1, \cdots, 4, \gamma_i = 2, k_i = 0.25, \sigma_i = 0.25$, and $\delta_i = 0.1, i = 1, \cdots, 10$. Solving the LMI (6.9) gives

$$P = \begin{bmatrix} 0.8660 & -0.5000 \\ -0.5000 & 0.8660 \end{bmatrix}.$$

Then, we further get that

$$K = \begin{bmatrix} -1.0000 & -1.7321 \end{bmatrix}, \quad \Gamma = \begin{bmatrix} 1.0000 & 1.7321 \\ 1.7321 & 3.0000 \end{bmatrix}.$$

The initial states of all agents are randomly chosen.

Under the designed adaptive dynamic event-based protocol (6.2)–(6.5), we can obtain the following simulation results. First, Fig. 6.2 represents the consensus errors $x_i - x_1, i = 2, \cdots, 10$, indicating that consensus can be achieved. With the asymptotic realization of consensus, the adaptive gains d_i and e_i gradually converge to stable values; see Figs. 6.3 and 6.4. Furthermore, we record the triggering instants in Fig. 6.5, which demonstrates that the events only happen at discrete instants. In other words, there is no need to carry out continuous communication or continuous updating.

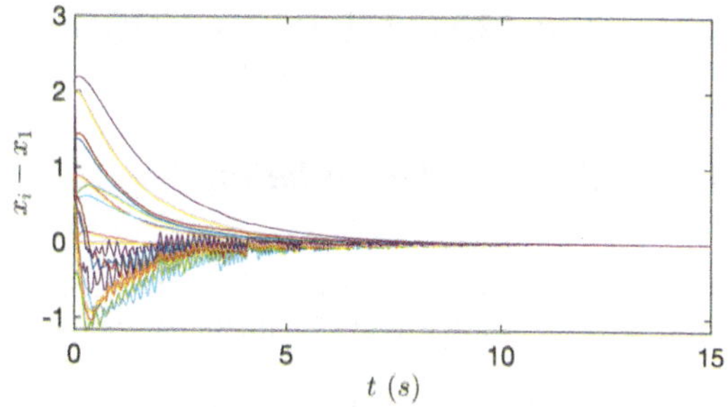

Fig. 6.2 The consensus errors $x_i - x_1, i = 2, \cdots, 10$

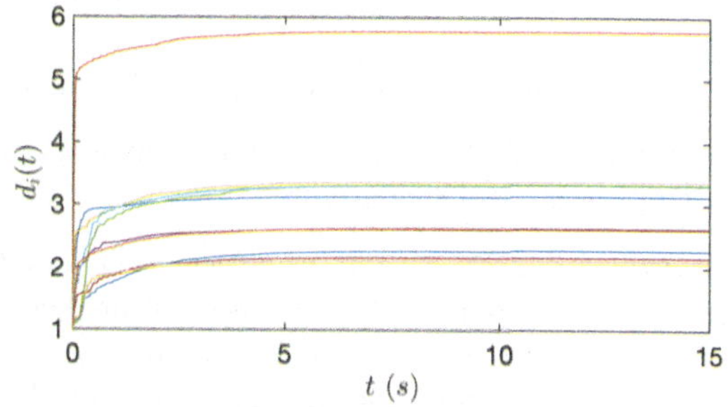

Fig. 6.3 The time-varying gains $d_i, i = 1, \cdots, 10$

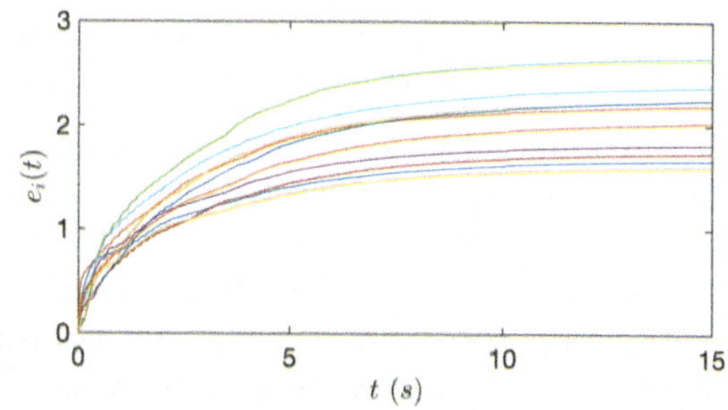

Fig. 6.4 The time-varying gains $e_i, i = 1, \cdots, 10$

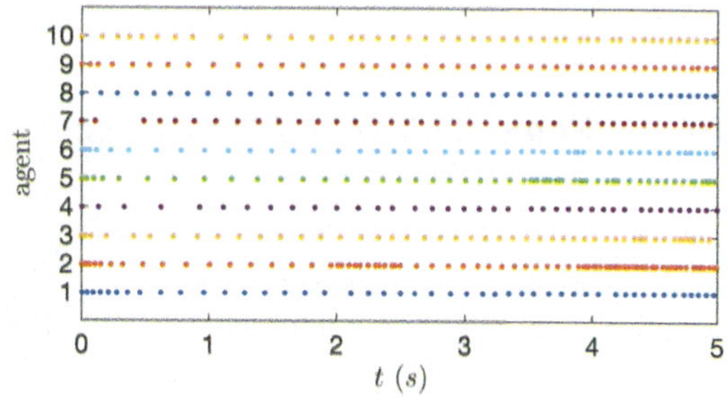

Fig. 6.5 Triggering instants of each agent t_k^i, $i = 1, \cdots, 10$

6.2 Robust Event-Triggered Tracking Control with Matching Uncertainties

6.2.1 Problem Formulation

In this section, we consider multi-agent systems in the presence of a leader and several followers. Particularly, the agent dynamics are given as

$$\begin{aligned} \dot{x}_0 &= Ax_0 + Bu_0, \\ \dot{x}_i &= Ax_i + B[u_i + f_i(t, x_i)], \; i = 1, \cdots, N. \end{aligned} \tag{6.34}$$

Throughout this section, we assume that the following assumption holds.

Assumption 6.1 The leader's control input u_0 is bounded, satisfying $\|u_0\| \leq \gamma$, where γ is a positive constant.

Assumption 6.1 is not strong, since unbounded control inputs are not implementable. The nonzero control input of the leader is often used to achieve certain objectives, such as to reach a desirable consensus value or to avoid hazardous obstacles. The leader-follower graph among the $N+1$ agents is denoted by $\mathcal{G}$, satisfies the following assumption.

Assumption 6.2 Graph $\mathcal{G}$ contains a directed spanning tree with the leader as a root and the subgraph associated with the followers is undirected.

Also assume that the leader receives no information from any follower and the state (but not the control input) of the leader is available to only a subset of the followers. The Laplacian matrix $\mathcal{L}$ associated with the leader-follower graph can be partitioned as $\mathcal{L} = \begin{bmatrix} 0 & 0_{1\times N} \\ \mathcal{L}_2 & \mathcal{L}_1 \end{bmatrix}$, where $\mathcal{L}_1 \in \mathbf{R}^{N\times N}$ is positive definite [53].

The objective of this section is to solve distributed tracking problem by designing distributed event-based protocols such that the followers can track the leader in the sense of $\lim_{t\to\infty} \|x_i(t) - x_0(t)\| = 0$, $i = 1, \cdots, N$, and by avoiding continuous communication and continuous control updating.

6.2.2 Robust Adaptive Event-Triggered Tracking Controllers

Denote $z = [z_1^T, \cdots, z_N^T]^T = [x_1^T, \cdots, x_N^T]^T - \mathbf{1}_N \otimes x_0$. Now, we define the tracking error as $\eta = (\mathcal{L}_1 \otimes I_n)z$, where $\eta = [\eta_1^T, \cdots, \eta_N^T]^T$. Then, we design an adaptive event-triggered controller for each follower as follows:

$$\begin{aligned} u_i(t) &= \hat{d}_i(t)K\hat{\eta}_i(t) + \hat{e}_i(t)g(K\hat{\eta}_i(t)), \\ \dot{d}_i(t) &= \hat{\eta}_i^T(t)\Gamma\hat{\eta}_i(t), \\ \dot{e}_i(t) &= \|K\hat{\eta}_i(t)\|, \; i = 1, \cdots, N, \end{aligned} \tag{6.35}$$

where $\hat{d}_i(t) = d_i(t_k^i)$, $\hat{e}_i(t) = e_i(t_k^i)$, $\forall t \in [t_k^i, t_{k+1}^i)$, $d_i(t)$ and $e_i(t)$ are adaptive coupling gains for agent i with initial values $d_i(0) \geq 1$ and $e_i(0) \geq 1$, $\hat{\eta}_i = \sum_{j=0}^N a_{ij}(\hat{x}_i - \hat{x}_j)$, and K and Γ are the feedback gain matrices.

To determine when to carry out communications and update controllers, we design the following dynamic triggering conditions:

$$
\begin{aligned}
t_{k+1}^0 &\triangleq \inf\{t > t_k^0 \mid \tilde{x}_0^T(t)\Gamma\tilde{x}_0(t) + \sqrt{\tilde{x}_0^T(t)\Gamma\tilde{x}_0(t)} \geq \gamma_0\varepsilon_0(t), \\
t_{k+1}^i &\triangleq \inf\{t > t_k^i \mid \tilde{x}_i^T(t)\Gamma\tilde{x}_i(t) + \sqrt{\tilde{x}_i^T(t)\Gamma\tilde{x}_i(t)} \geq \gamma_i\varepsilon_i(t) \\
&\quad \vee\ |\tilde{d}_i(t)| \geq \theta_1 e^{-\theta_2(t-t_k^i)} \ \vee\ |\tilde{e}_i(t)| \geq \theta_3 e^{-\theta_4(t-t_k^i)}\},\ i = 1, \cdots, N,
\end{aligned}
\tag{6.36}
$$

where $\tilde{x}_0(t) = x_0(t_k^0) - x_0(t)$, $\gamma_i > 0$, $\theta_i, i = 1, \cdots, 4$, are positive constants, and $\varepsilon_i(t)$ satisfies the following dynamics:

$$
\dot{\varepsilon}_i(t) = -k_i\varepsilon_i(t) - \sigma_i\tilde{x}_i^T(t)\Gamma\tilde{x}_i(t) - \delta_i\sqrt{\tilde{x}_i^T(t)\Gamma\tilde{x}_i(t)},\ i = 0, \cdots, N,
\tag{6.37}
$$

with $\varepsilon_i(0) > 0$, $k_i > 0$, $\sigma_i > 0$, and $\delta_i > 0$.

Remark 6.5 Different from the leaderless consensus control problem considered in the previous section, for the distributed tracking problem considered in this section, we have designed different event-triggering conditions for the leader and the followers, respectively. Note that due to the decoupling feature of the triggering conditions, the internal variables for the leader and the followers are the same. The difference is that the triggering conditions for the followers contain the variables $\tilde{d}_i$ and $\tilde{e}_i$, while the leader does not have such variables.

Letting $\tilde{\eta}_i = \sum_{j=1}^N a_{ij}(\tilde{x}_i - \tilde{x}_j)$, we have $\tilde{\eta}_i = \hat{\eta}_i - \eta_i$. Denote $\bar{d}_i$ and $\bar{e}_i$ as the virtual adaptive gains, whose dynamics are given by

$$
\begin{aligned}
\dot{\bar{d}}_i &= \eta_i^T\Gamma\eta_i, \\
\dot{\bar{e}}_i &= \|K\eta_i\|,
\end{aligned}
\tag{6.38}
$$

with initial value $\bar{d}_i(0) \geq 1$ and $\bar{e}_i(0) \geq 0$, $i = 1, \cdots, N$. Further let $\check{d}_i = d_i - \bar{d}_i$ and $\check{e}_i = e_i - \bar{e}_i$. We can derive the closed-loop dynamics of z as

$$
\begin{aligned}
\dot{z} = {} & (I_N \otimes A)z + (\bar{D} \otimes BK)\eta + (D \otimes BK)\tilde{\eta} + (\check{D} \otimes BK)\eta + (\tilde{D} \otimes BK)\hat{\eta} \\
& + (wI_N \otimes B)G' + (I_N \otimes B)F - \mathbf{1}_N \otimes Bu_0,
\end{aligned}
\tag{6.39}
$$

where $\eta = [\eta_1^T, \cdots, \eta_N^T]^T$, $\tilde{\eta} = [\tilde{\eta}_1^T, \cdots, \tilde{\eta}_N^T]^T$, $\hat{\eta} = [\hat{\eta}_1^T, \cdots, \hat{\eta}_N^T]^T$, $\bar{D} = \text{diag}\{\bar{d}_1, \cdots, \bar{d}_N\}$, $D = \text{diag}\{d_1, \cdots, d_N\}$, $\tilde{D} = \text{diag}\{\tilde{d}_1, \cdots, \tilde{d}_N\}$, $\check{D} =$

$\text{diag}\{\check{d}_1, \cdots, \check{d}_N\}$, $G' = [g^T(K\hat{\eta}_1), \cdots, g^T(K\hat{\eta}_N)]^T$, and $F = [f_1^T, \cdots, f_N^T]^T$. Then, we can obtain the dynamics of η as follows:

$$\begin{aligned}\dot{\eta} &= (I_N \otimes A)\eta + (\mathcal{L}_1 \bar{D} \otimes BK)\eta + (\mathcal{L}_1 D \otimes BK)\tilde{\eta} + (\mathcal{L}_1 \check{D} \otimes BK)\eta \\ &\quad + (\mathcal{L}_1 \tilde{D} \otimes BK)\hat{\eta} + (\mathcal{L}_1 \check{E} \otimes B)G' + (\mathcal{L}_1 \otimes B)F - \mathcal{L}_1 \mathbf{1}_N \otimes Bu_0.\end{aligned} \tag{6.40}$$

Now, we are ready to show the convergence of the tracking error η.

Theorem 6.3 *Assuming that Assumption 6.2 holds, the followers can track the leader under the adaptive event-triggered protocol (6.35)–(6.37) whose parameters are chosen as in Theorem 6.1. Moreover, there does not exist Zeno behavior.*

Proof Choose a Lyapunov function candidate as

$$V_2 = z^T(\mathcal{L}_1 \otimes Q)z + \sum_{i=1}^{N} \left[\frac{1}{2}(\bar{d}_i - \alpha_d')^2 + \beta_d' \bar{d}_i V_0' + \frac{1}{2}(\bar{e}_i - \alpha_e')^2 + \beta_e' \bar{e}_i V_0'\right], \tag{6.41}$$

where α_d', α_e', β_d', and β_e' are positive constants to be determined, and $V_0' = \sum_{j=0}^{N} \frac{\varepsilon_j}{k_j}$. The time derivative of V_2 along the trajectory of (6.39) is given by

$$\begin{aligned}\dot{V}_2 &= z^T[\mathcal{L}_1 \otimes (QA + A^T Q)]z - 2\eta^T(\bar{D} \otimes \Gamma)\eta \\ &\quad - 2\eta^T(D \otimes \Gamma)\tilde{\eta} - 2\eta^T(\check{D} \otimes \Gamma)\eta - 2\eta^T(\tilde{D} \otimes \Gamma)\hat{\eta} \\ &\quad + 2\eta^T(\check{E} \otimes QB)G' + 2\eta^T(I_N \otimes QB)F - 2\eta^T(I_N \otimes QBu_0) \\ &\quad + \sum_{i=1}^{N}(\bar{d}_i - \alpha_d' + \beta_d' V_0')\eta_i^T \Gamma \eta_i + \sum_{i=1}^{N}(\bar{e}_i - \alpha_e' + \beta_e' V_0')\|K\eta_i\| \\ &\quad - \sum_{i=1}^{N}(\beta_d' \bar{d}_i + \beta_e' \bar{e}_i) \sum_{j=0}^{N}\left(\varepsilon_j + \frac{\sigma_j}{k_j}\tilde{x}_j^T \Gamma \tilde{x}_j + \frac{\delta_j}{k_j}\sqrt{\tilde{x}_j^T \Gamma \tilde{x}_j}\right).\end{aligned} \tag{6.42}$$

Similarly as in the proof of Theorem 6.1, we have

$$\begin{aligned}-2\eta^T(D \otimes \Gamma)\tilde{\eta} &\le \frac{1}{2}\eta^T(\bar{D} \otimes \Gamma)\eta + \frac{1}{2}\eta^T(\check{D} \otimes \Gamma)\eta \\ &\quad + \sum_{i=1}^{N}\left(8\bar{d}_i + 2\beta_1'\right) l_{ii}^2 \sum_{j=0}^{N} \tilde{x}_j^T \Gamma \tilde{x}_j,\end{aligned} \tag{6.43a}$$

$$
\begin{aligned}
-\frac{3}{2}\eta^T(\check{D}\otimes\Gamma)\eta &= \sum_{i=1}^{N} -\frac{3}{2}\check{d}_i\eta_i^T\Gamma\eta_i \\
&\leq \sum_{i=1}^{N}[\frac{1}{2}\bar{d}_i(t) - \frac{3}{2}(\check{d}_i(0) + \frac{1}{3}\bar{d}_i(0)) \\
&\quad + 6\sum_{i=1}^{N} l_{ii}^2 \sum_{j=0}^{N} \frac{\gamma_i\varepsilon_j(0)}{k_j}]\eta_i^T\Gamma\eta_i \\
&\leq \frac{1}{2}\eta^T(\bar{D}\otimes\Gamma)\eta + \eta^T(\beta_2' I_N\otimes\Gamma)\eta,
\end{aligned}
\tag{6.43b}
$$

$$
\begin{aligned}
-2\eta^T(\tilde{D}\otimes\Gamma)\hat{\eta} &= -2\sum_{i=1}^{N}\tilde{d}_i\eta^T\Gamma\hat{\eta}_i \\
&\leq \theta_1\sum_{i=1}^{N}\eta_i^T\Gamma\eta_i + \theta_1\sum_{i=1}^{N}\hat{\eta}_i^T\Gamma\hat{\eta}_i \\
&\leq 3\theta_1\sum_{i=1}^{N}\eta_i^T\Gamma\eta_i + 2\theta_1\sum_{i=1}^{N}\tilde{\eta}_i^T\Gamma\tilde{\eta}_i \\
&\leq 3\theta_1\sum_{i=1}^{N}\eta_i^T\Gamma\eta_i + 4\theta_1\sum_{i=1}^{N} l_{ii}^2\sum_{j=0}^{N}\tilde{x}_j^T\Gamma\tilde{x}_j,
\end{aligned}
\tag{6.43c}
$$

$$
\begin{aligned}
2\eta^T(\hat{E}\otimes QB)G' &\leq -\sum_{i=1}^{N}\bar{e}_i(t)\|B^T Q\eta_i\| + \sum_{i=1}^{N}\beta_3'\|B^T Q\eta_i\| \\
&\quad + \sum_{i=1}^{N}\beta_4'\bar{e}_i l_{ii}\sum_{j=0}^{N}\sqrt{\tilde{x}_j\Gamma\tilde{x}_j},
\end{aligned}
\tag{6.43d}
$$

$$
\begin{aligned}
2\eta^T(I_N\otimes QB)F &\leq 2\sum_{i=1}^{N}\|B^T Q\eta_i\|\cdot\|f_i\| \\
&\leq 2f_0\sum_{i=1}^{N}\|B^T Q\eta_i\|,
\end{aligned}
\tag{6.43e}
$$

$$
\begin{aligned}
-2\eta^T(I_N \otimes QBu_0) &\leq 2\sum_{i=1}^{N} \|B^T Q\eta_i\| \cdot \|u_0\| \\
&\leq 2\gamma \sum_{i=1}^{N} \|B^T Q\eta_i\|,
\end{aligned}
\tag{6.43f}
$$

where $\beta_1' \geq \max\{-2\bar{d}_i(0) + 2\check{d}_i(0) + 8l_{ii}^2 \sum_{j=0}^{N} \frac{\gamma_j \varepsilon_j(0)}{k_j}\}$, $\beta_2' \geq \max\{-\frac{3}{2}(\check{d}_i(0) + \frac{1}{3}\bar{d}_i(0)) + 6\sum_{i=1}^{N} l_{ii}^2 \sum_{j=0}^{N} \frac{\gamma_i \varepsilon_j(0)}{k_j}\}$, $\beta_3' \geq \max\{2\theta_3 - \bar{e}_i(0) - 2\check{e}_i(0) + \sqrt{2}l_{ii} \sum_{j=0}^{N} \sqrt{\frac{\gamma_j \varepsilon_j(0)}{k_j}}\}$, and $\beta_4' = \max\{4(1 + \theta_3 + \check{e}_i(0) + \sqrt{2}l_{ii} \sum_{j=0}^{N} \sqrt{\frac{\gamma_j \varepsilon_j(0)}{k_j}})\}$.

According to (6.43) and by choosing $\beta_d' \geq 2(4 + \beta_1 + 2\theta_1') \max\{l_{ii}^2\} \max\{\frac{k_i}{\sigma_i}\}$, $\beta_e' \geq \beta_4' \max\{l_{ii}\} \max\{\frac{k_i}{\delta_i}\}$, $\alpha_d' = \alpha_1' + \beta_d' V_0(0) + \beta_2' + 3\theta_1$, and $\alpha_e' = \beta_e' V_0(0) + \beta_3' + 2f_0 + 2\gamma$, we can rewrite (6.42) as follows:

$$
\begin{aligned}
\dot{V}_2 &\leq z^T[\mathcal{L}_1 \otimes (QA + A^T Q)]z - \eta^T(\alpha_1' I_N \otimes \Gamma)\eta \\
&\leq z^T[\mathcal{L}_1 \otimes (QA + A^T Q - \Gamma)]z \\
&\leq \frac{1}{\Gamma_N(\mathcal{L}_1)} \eta^T[I_N \otimes (QA + A^T Q - \Gamma)]\eta \\
&\leq 0,
\end{aligned}
\tag{6.44}
$$

where the second inequality is obtained by applying Lemma 2.2 and by choosing $\alpha_1' \geq \frac{1}{\Gamma_1(\mathcal{L}_1)}$, $\Gamma_1(\mathcal{L}_1)$ and $\Gamma_N(\mathcal{L}_1)$ denote the smallest and largest eigenvalues of $\mathcal{L}_1$, respectively.

Then, similar to the proof of Theorem 6.1, it follows from Lemma 2.4 that $\eta \to 0$ as $t \to \infty$, i.e., consensus is achieved. We can further exclude the Zeno behavior by following similar steps in the proof of Theorem 6.2. The details are omitted here for brevity. □

6.2.3 Simulation Results

Referring to [110], we consider a multiple air-vehicle system satisfying (6.1), where

$$
A = \begin{bmatrix} 0 & 1 & 0 & 0 \\ 0 & -5 & 0 & 0 \\ 5 & 0 & -5 & 0 \\ 0 & 5 & 0 & -5 \end{bmatrix}, \quad B = \begin{bmatrix} 0 \\ 1 \\ 0 \\ 0 \end{bmatrix}, \quad f_i(t, x_i) = i/10 \sin(t), \ i = 1, \cdots, 4,
$$

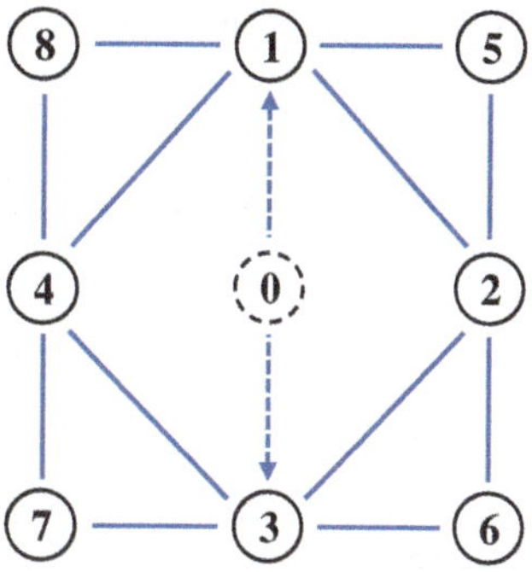

Fig. 6.6 A leader-follower communication graph

and $f_i(t, x_i)$, $i = 5, \cdots, 8$, obey normal distribution, respectively. The agents are connected by a network topology with 1 leader and 8 followers, as given in Fig. 6.6.

To track the leader's state and cover the model uncertainties, we use the adaptive dynamic event-triggered protocol composed of (6.35)–(6.37). The parameters in the protocol are chosen as $\theta_i = 1$, $i = 1, \cdots, 4$, $\gamma_i = 2$, $k_i = 0.25$, $\sigma_i = 1$, and $\delta_i = 0.2$, $i = 0, 1, \cdots, 8$. Solving the LMI (6.9) gives

$$P = \begin{bmatrix} 0.4066 & -1.4688 & -0.2585 & 0.7403 \\ -1.4688 & 7.8830 & 0.6929 & -3.8795 \\ -0.2585 & 0.6929 & 10.1878 & -0.3536 \\ 0.7403 & -3.8795 & -0.3536 & 11.9334 \end{bmatrix}.$$

Then, we can get that

$$K = \begin{bmatrix} -1.4049 & -0.4108 & -0.0093 & -0.0467 \end{bmatrix},$$

$$\Gamma = \begin{bmatrix} 1.9737 & 0.5771 & 0.0131 & 0.0656 \\ 0.5771 & 0.1687 & 0.0038 & 0.0192 \\ 0.0131 & 0.0038 & 0.0001 & 0.0004 \\ 0.0656 & 0.0192 & 0.0004 & 0.0022 \end{bmatrix}.$$

The simulation results are shown as follows. Figure 6.7 indicates that the tracking errors $x_i - x_0$, $i = 1, \cdots, 8$ asymptotically converge to 0. It follows from Figs. 6.8 and 6.9 that the time-varying gains d_i and e_i tend to bounded constants. Events only happen at discrete instants as shown in Fig. 6.10.

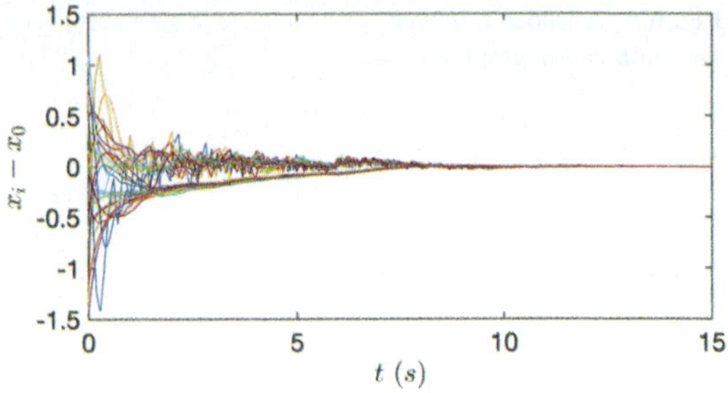

Fig. 6.7 The tracking errors $x_i - x_0, i = 1, \cdots, 8$

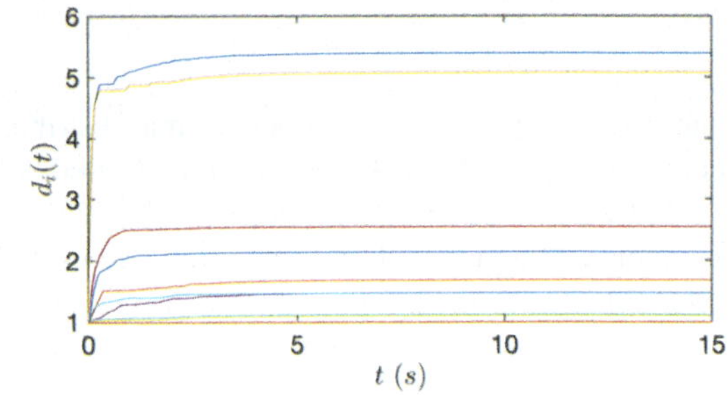

Fig. 6.8 The time-varying gains $d_i, i = 1, \cdots, 8$

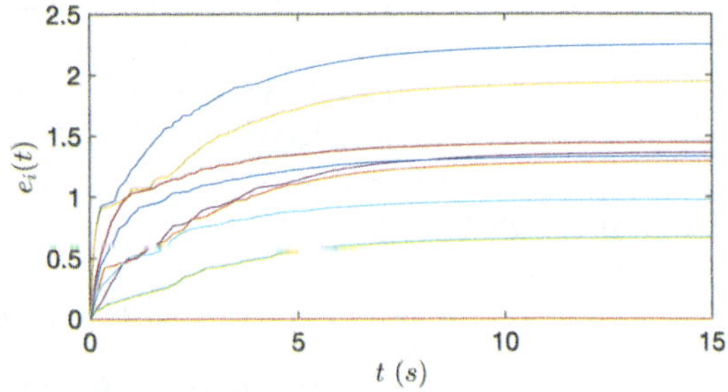

Fig. 6.9 The time-varying gains $e_i, i = 1, \cdots, 8$

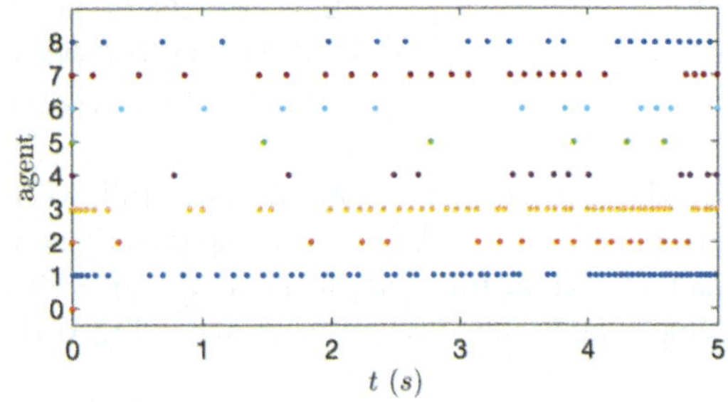

Fig. 6.10 Triggering instants of each agent t_k^i, $i = 0, 1, \cdots, 8$

6.3 Distributed Event-Triggered Output Regulation

6.3.1 Problem Formulation

In this section, consider a group of N heterogeneous linear agents, whose dynamics are described by

$$\begin{aligned} \dot{x}_i &= A_i x_i + B_i u_i + E_i v, \\ e_i &= C_i x_i + D_i v, \ i = 1, \cdots, N, \end{aligned} \tag{6.45}$$

where $x_i \in \mathbf{R}^{n_i}$, $u_i \in \mathbf{R}^{m_i}$, and $e_i \in \mathbf{R}^{p_i}$ denote the state, the control input, and the regulated output of agent i, respectively, $v \in \mathbf{R}^q$ denotes the state of exogenous signal which can be either the disturbance to be rejected or a reference signal to be tracked. The exogenous signal v can be generated by the following exosystem:

$$\dot{v} = Sv, \tag{6.46}$$

where $S \in \mathbf{R}^{q \times q}$.

Here, we need the following standard assumptions [42].

Assumption 6.3 The pair (A_i, B_i) is stabilizable.

Assumption 6.4 The matrix S has no eigenvalues with negative real parts.

Assumption 6.5 For all $\Gamma \in \sigma(S)$, in which $\sigma(S)$ is the spectrum of S, the condition $\mathrm{rank}\left(\begin{bmatrix} A_i - \Gamma I & B_i \\ C_i & 0 \end{bmatrix}\right) = n_i + p_i$ holds.

Assumption 6.6 There exist solution pairs $(U_i \in \mathbf{R}^{m_i \times q}, H_i \in \mathbf{R}^{n_i \times q})$ for the following equations:

$$\begin{aligned} A_i H_i + B_i U_i + E_i &= H_i S, \\ C_i H_i + D_i &= 0, \ i = 1, \cdots, N. \end{aligned} \tag{6.47}$$

In this section, we call the exosystem described by (6.46) the leader and the N subsystems described by (6.45) followers. The network topology among the $N + 1$ agents can be modeled by a directed graph $\mathcal{G}$, and satisfies the following assumption.

Assumption 6.7 The graph $\mathcal{G}$ contains a directed spanning tree with the root node being the leader.

Compared to Assumption 6.2 in the previous section, the above assumption is more relaxed, since the subgraph among followers can be directed. The Laplacian matrix $\mathcal{L}$ can be partitioned as $\mathcal{L} = \begin{bmatrix} 0 & 0_{1 \times N} \\ \mathcal{L}_2 & \mathcal{L}_1 \end{bmatrix}$, where $\mathcal{L}_1 \in \mathbf{R}^{N \times N}$ and $\mathcal{L}_2 \in \mathbf{R}^{N \times 1}$. It is not difficult to see that $\mathcal{L}_1$ is a nonsingular M-matrix [68].

The aim of this section is to develop appropriate distributed adaptive event-triggered controller to solve the cooperative output regulation problem. In particular, (1) the overall closed-loop system is asymptotically stable when the exogenous signal $v = 0$; (2) for any initial conditions, $\lim_{t\to\infty} e_i(t) = 0$; and (3) there does not exist Zeno behavior.

6.3.2 Adaptive Event-Triggered Output Regulation Controllers

To be convenient, we denote $v_0 \triangleq v$. To estimate the state v, following the ideas in Chap. 4, we construct the following distributed adaptive event-triggered observer for each follower:

$$\begin{aligned} \dot{v}_i(t) &= S v_i(t) + (\hat{d}_i(t) + \rho_i(t))\mu K \hat{\eta}_i(t), \\ \dot{d}_i(t) &= \hat{\eta}_i^T(t)\Gamma\hat{\eta}_i(t),\ i = 1, \cdots, N, \end{aligned} \tag{6.48}$$

where μ is an arbitrary positive constant, v_i is the estimate of v, K and Γ are the feedback gain matrices, $\rho_i(t) = \hat{\eta}_i^T P \hat{\eta}_i$, $d_i(t)$ is the adaptive coupling gain with $d_i(0) \geq 1$, $\hat{d}_i(t) = d_i(t_k^i)$, $\hat{\eta}_i(t) = \sum_{j=0}^N a_{ij}(\hat{v}_i(t) - \hat{v}_j(t))$, and $\hat{v}_i(t) = v_i(t_k^i)$, $\forall t \in [t_k^i, t_{k+1}^i)$, where $t \in [t_k^i, t_{k+1}^i)$ denotes the calculated time between the k-th and the $(k+1)$-th triggering time instants.

Let $\tilde{d}_i = \hat{d}_i - d_i$, $i = 1, \cdots, N$, and $\tilde{v}_i = \hat{v}_i - v_i$, $i = 0, 1, \cdots, N$. The triggering functions for each agent are designed as follows:

$$\begin{aligned} f_i(t) &= \tilde{v}_i^T(t)\tilde{v}_i(t) - \alpha_i\epsilon_i(t), \\ \hat{f}_i(t) &= |\tilde{d}_i| - \beta_1 e^{-\gamma_1(t - t_k^i)}, \\ \tilde{f}_0(t) &= \tilde{v}_0^T(t)\tilde{v}_0(t) - \alpha_0\epsilon_0(t), \end{aligned} \tag{6.49}$$

where $\alpha_i > 0$, β_1 and γ_1 are the positive constants, $\dot{\epsilon}_i = -k_i\epsilon_i - \sigma_i\tilde{v}_i^T\tilde{v}_i$, $i = 0, 1, \cdots, N$, where $\epsilon_i(0) > 0$, $k_i > 0$, and $\sigma_i > 0$. At event instants of the leader when $\tilde{f}_0(t) \geq 0$, the leader will send its current state information to the informed followers that can directly receive information from the leader. At event instants of the followers when $f_i(t) \geq 0$ or $\hat{f}_i(t) \geq 0$, they will update their controllers and send their state information to their out-neighbors.

Let $\eta = [\eta_1^T, \cdots, \eta_N^T]^T$ and $\hat{\eta} = [\hat{\eta}_1^T, \cdots, \hat{\eta}_N^T]^T$, where $\eta_i(t) = \sum_{j=0}^N a_{ij}(v_i(t) - v_j(t))$. Then, we have

$$\begin{aligned} \dot{\eta}_i(t) &= S\eta_i(t) + (\hat{d}_i(t) + \rho_i(t))\mu K\hat{\eta}_i(t), \\ \dot{d}_i(t) &= \hat{\eta}_i^T(t)\Gamma\hat{\eta}_i(t),\ i = 1, \cdots, N. \end{aligned}$$

Define $\bar{d}_i(t)$ and $\bar{\rho}_i(t)$ as the virtual adaptive gains with dynamics that $\dot{\bar{d}}_i(t) = \eta_i^T(t)\Gamma\eta_i(t)$ with $\bar{d}_i(0) \geq 1$ and $\bar{\rho}_i(t) = \eta_i^T(t)P\eta_i(t)$ with $\bar{\rho}_i(0) \geq 1$, $i = 1, \cdots, N$, respectively. Let $\check{d}_i = d_i - \bar{d}_i$, $\check{\rho}_i = \rho_i - \bar{\rho}_i$, $v_s = [v_1^T, \cdots, v_N^T]^T$, $\tilde{\eta}_s = \hat{\eta}_s - \eta_s$, $\eta_s = [\eta_1^T, \cdots, \eta_N^T]^T$, $\tilde{\eta}_s = [\tilde{\eta}_1^T, \cdots, \tilde{\eta}_N^T]^T$, and $\hat{\eta}_s = [\hat{\eta}_1^T, \cdots, \hat{\eta}_N^T]^T$. Then, it is easy to get the consensus error $\eta_s = (\mathcal{L}_1 \otimes I_q)(v_s - \mathbf{1}_N \otimes v_0)$. The dynamics of η_s are described by

$$\dot{\eta}_s = (I_N \otimes S)\eta_s + [\mu\mathcal{L}_1(\bar{D}_s + \bar{\rho}_s) \otimes K]\eta_s + [\mu\mathcal{L}_1(\check{D}_s + \check{\rho}_s) \otimes K]\eta_s$$
$$+ [\mu\mathcal{L}_1(D_s + \rho_s) \otimes K]\tilde{\eta}_s + [\mu\mathcal{L}_1\tilde{D}_s \otimes K]\hat{\eta}_s,$$

where $D_s = \text{diag}\{d_1, \cdots, d_N\}$, $\check{D}_s = \text{diag}\{\check{d}_1, \cdots, \check{d}_N\}$, $\bar{D}_s = \text{diag}\{\bar{d}_1, \cdots, \bar{d}_N\}$, $\tilde{D}_s = \text{diag}\{\tilde{d}_1, \cdots, \tilde{d}_N\}$, $\rho_s = \text{diag}\{\rho_1, \cdots, \rho_N\}$, $\check{\rho}_s = \text{diag}\{\check{\rho}_1, \cdots, \check{\rho}_N\}$, and $\bar{\rho}_s = \text{diag}\{\bar{\rho}_1, \cdots, \bar{\rho}_N\}$.

As a special case of Theorem 4.3, we can get the following result.

Theorem 6.4 *Suppose that Assumptions 6.3–6.7 hold. The event-triggered observers composed of* (6.48) *and* (6.49) *can track the state of the exosystem* (6.46) *and the coupling gains d_i in* (6.48) *converge to some finite steady-state values, if $K = -P$ and $\Gamma = P^2$, where $P > 0$ is a solution to the ARE:*

$$PS - P^2 + S^T P + I = 0. \tag{6.50}$$

Furthermore, the Zeno behavior can be excluded.

According to the estimate value v_i , an observer-based output feedback controller can be designed as follows:

$$u_i = J_i x_i + \hat{J}_i v_i,\ i = 1, \cdots, N, \tag{6.51}$$

where $J_i \in \mathbf{R}^{m_i \times n_i}$ and $\hat{J}_i \in \mathbf{R}^{m_i \times q}$ are the feedback gain matrices.

Theorem 6.5 *Suppose that Assumptions 6.3–6.7 hold. Choose J_i such that $A_i + B_i J_i$ are Hurwitz, and choose $\hat{J}_i = U_i - J_i H_i$. The cooperative output regulation problem can be tackled under the protocol composed of the triggering function* (6.49)*, the distributed controller* (6.51) *and the adaptive observer* (6.48).

Proof Letting $\iota_i = x_i - H_i v_i, i = 1, \cdots, N$, and substituting (6.45) and (6.48) into ι_i yield

$$\dot{\iota}_i = (A_i + B_i J_i)\iota_i + (A_i H_i + B_i U_i - H_i S)v_i$$
$$+ B_i(\hat{J}_i - U_i + J_i H_i)v_i - H_i(\hat{d}_i + \hat{\rho}_i)\mu\hat{\eta}_i(t).$$

It follows from Theorem 6.4 that $\lim_{t\to\infty}\hat{\eta}_i(t) = 0$. By choosing $\hat{J}_i = U_i - J_i H_i$ and invoking (6.47), we can get that

$$\dot{\iota}_i = (A_i + B_i J_i)\iota_i.$$

We can observe that $\lim_{t\to\infty}\iota_i(t) = 0$, implying that $\lim_{t\to\infty}(x_i - H_i v_i) = 0$. Noting that $0 = C_i H_i + D_i$, we can derive that

$$\lim_{t\to\infty} e_i(t) = \lim_{t\to\infty}(C_i x_i + D_i v) = \lim_{t\to\infty}(C_i H_i + D_i)v = 0.$$

Then, we can obtain that $\lim_{t\to\infty} x_i(t) = 0$ when $v = 0$, which implies that $\lim_{t\to\infty} e_i(t) = 0$. Therefore, the cooperative output regulation problem is solved. □

6.3.3 Simulation Results

Consider a network of five agents, described by (6.45) with

$$A_i = \begin{bmatrix} 0 & 1 & 0 \\ 0 & 0 & 1 \\ 0 & a_i^1 & a_i^2 \end{bmatrix}, B_i = \begin{bmatrix} 0 \\ 0 \\ a_i^3 \end{bmatrix}, E_i = \begin{bmatrix} -1 & 0 \\ -2 & 0 \\ a_i^4 & a_i^5 \end{bmatrix},$$

$$C_i = \begin{bmatrix} a_i^6 & a_i^7 & a_i^8 \end{bmatrix}, D_i = \begin{bmatrix} a_i^9 & a_i^{10} \end{bmatrix}, S = \begin{bmatrix} 0 & 1 \\ -1 & 0 \end{bmatrix}, i = 1, \cdots, 4,$$

where $a_1^1 = a_2^1 = 0$, $a_3^1 = -10$, $a_4^1 = -1$, $a_1^2 = -1$, $a_2^2 = -2$, $a_3^2 = -3$, $a_4^2 = -4$, $a_1^3 = a_3^3 = a_4^3 = 1$, $a_2^3 = 2$, $a_1^4 = -1$, $a_2^4 = -2$, $a_3^4 = -1$, $a_4^4 = -2$, $a_1^5 = -2$, $a_2^5 = -5$, $a_3^5 = 8$, $a_4^5 = -1$, $a_1^6 = a_4^6 = 1$, $a_2^6 = 2$, $a_3^6 = -2.7$, $a_1^7 = a_4^7 = 0.5$, $a_2^7 = -2.7$, $a_3^7 = 6.4$, $a_1^8 = a_4^8 = 2$, $a_2^8 = 3.6$, $a_3^8 = 0.7$, $a_1^9 = a_4^9 = -3$, $a_2^9 = -5.6$, $a_3^9 = 2$, $a_1^{10} = a_4^{10} = -1.5$, $a_2^{10} = 0.7$, and $a_3^{10} = -3.7$. Figure 6.11 depicts the communication links between the agents.

Solving the ARE (6.50) gives a solution as follows:

$$P = \begin{bmatrix} 1.4142 & -2.0483 \\ -6.6613 & 1.4142 \end{bmatrix}.$$

Then, we can get that

$$\Gamma = \begin{bmatrix} 0.5000 & 1.4484 \\ 4.7103 & 0.5000 \end{bmatrix}.$$

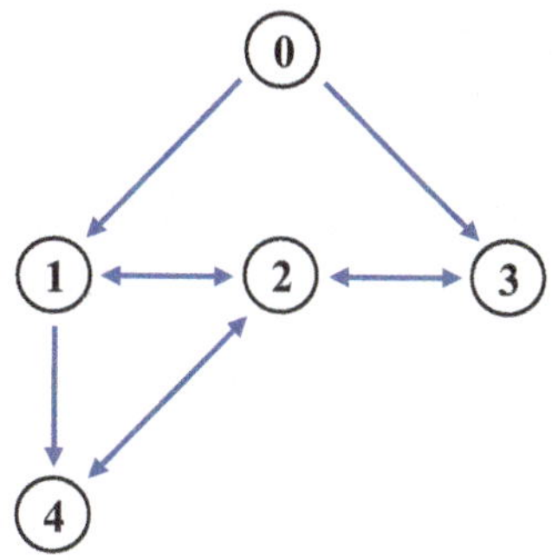

Fig. 6.11 A directed graph

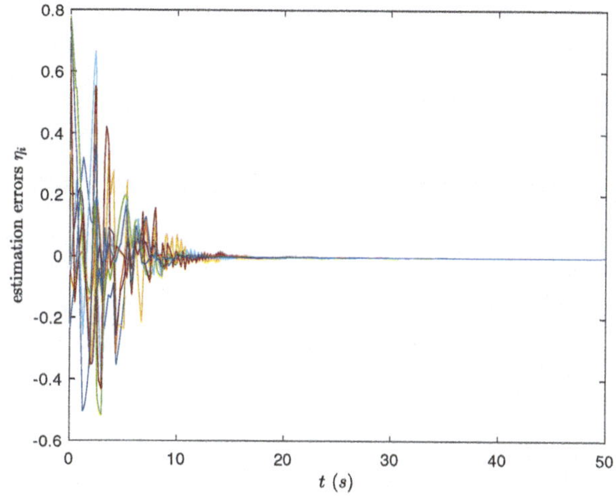

Fig. 6.12 Estimation errors η_i, $i = 1, \cdots, 4$

The gain matrices are set as $J_i = \begin{bmatrix} -1 & -1 & -1 \end{bmatrix}$ and $\hat{J}_i = \begin{bmatrix} 4 & 5 \end{bmatrix}$. Other parameters of (6.48) and (6.49) are chosen as $\mu = 1$, $\alpha_i = 2$, $\beta_1 = 1$, $\gamma_1 = 1$, $k_i = 0.25$, $\sigma_i = 0.25$, and $\epsilon_i(0) = 0.4$, for $i = 1, \cdots, 4$.

The evolution of the estimation errors η_i, $i = 1, \cdots, 4$, is shown in Fig. 6.12, from which it can be observed that all agents can estimate the state of the exosystem even if some of them cannot obtain the output information directly. The triggering instants for all agents are shown in Fig. 6.13, which indicates that the Zeno behavior is excluded. Figure 6.14 shows the regulated outputs e_i, $i = 1, \cdots, 4$ for t from 0 s to 50 s, implying that the regulated outputs asymptotically converge to zero.

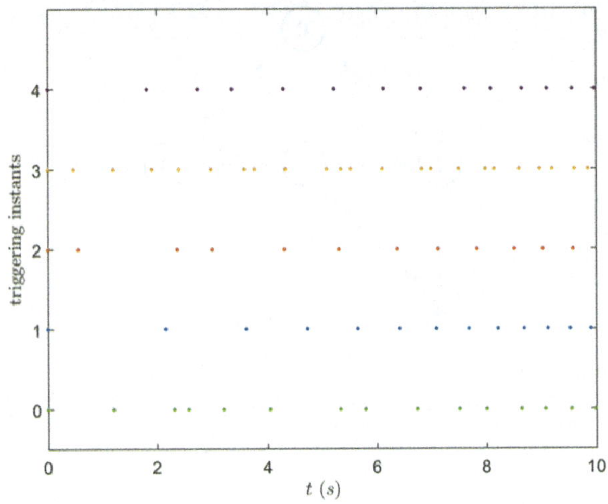

Fig. 6.13 Triggering instants of the agents

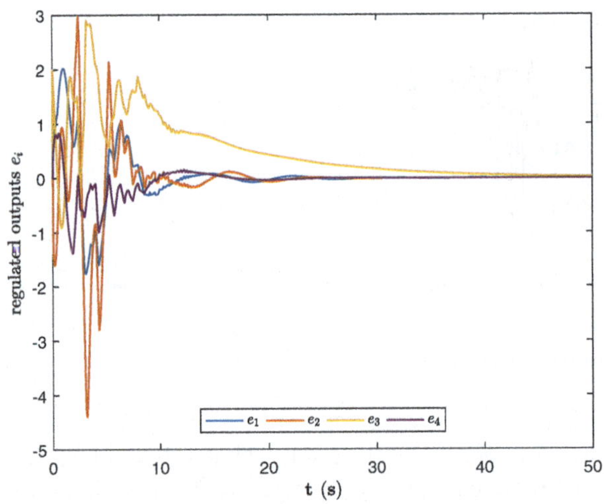

Fig. 6.14 Regulated outputs $e_i, i = 1, \cdots, 4$

6.4 Notes

The results of this chapter are mainly based on [10, 12, 13, 15, 88]. We have used the event-triggered control, adaptive control, and sliding-mode control techniques to well solve the fully distributed event-based consensus and tracking problems for multi-agent systems in presence of matched uncertainties, with/without a dynamic leader. We also extend the results to output regulation control problem for a class of heterogeneous linear multi-agent systems. The results of this chapter are essentially different from [39, 40, 67, 82], which considered the event-triggered cooperative output regulation problem for undirected graphs.

Chapter 7
Distributed Robust Event-Triggered Control with Frequency-Domain Uncertainties

In Chap. 6, we have studied the robustness of event-triggered control in the presence of matched nonlinear uncertainties or uncertainties generated by exosystems from the perspectives of nonlinear control such as sliding mode control and output regulation. Actually, the robustness of event-triggered control with respect to norm-bounded uncertainties can be investigated in the framework of linear robust control [118]. Generally speaking, norm-bounded uncertainties consist of parameter uncertainties in the time domain and dynamic uncertainties in the frequency domain. The time-domain parameter uncertainties are relatively easier to handle, since the event-triggering mechanisms are also described in the time domain. Related works along this line include [41, 86, 93]. Apart from time-domain uncertainties, the frequency-domain uncertainties, including unmodeled dynamics, and modeling errors, are a more general class of uncertainties [118], which may be encountered and need to be dealt with in the event-triggered control problem.

Even though distributed control algorithms with continuous and ideal communications have been proved robust to various kinds of frequency-domain uncertainties such as additive dynamic uncertainties [57, 92], network multiplicative uncertainties [51, 52, 109], coprime factor uncertainties [46], yet the robustness of distributed event-triggered algorithms against frequency-domain uncertainties appears much more challenging. The reason lies in that the event-triggered algorithms belong to a special branch of the aperiodic sampling algorithms, whose definition, modeling, and methodology are all based on the time-domain analysis. More specifically, the triggering function, which decides whether the agent updates its state estimation and broadcasts it to its neighbors, is expressed in a time-domain form. In other words, there exists a gap between the time-domain event triggering and the frequency-domain uncertainties.

In this chapter, we intend to study the robustness of the event-triggered consensus algorithms against frequency-domain uncertainties. In Sect. 7.1, it is shown that the sampling errors are images of the linear operators acting on the consensus error of the sampled states. These linear operators are neither generally rational transfer

Z. Li et al., *Distributed Event-triggered Control*,
https://doi.org/10.1007/978-981-99-8170-0_7

matrices in $\mathcal{RH}_\infty$ nor sector-bounded uncertainties of logarithmic quantizers as in [26]. Nevertheless, we can ascertain from the triggering function that the operators are finite gain $\mathcal{L}_2$ stable, whose gain, depending on the sampling parameters and the Laplacian matrix of the topology graph, characterizes the extent of the sampling error introduced by the event triggering. In this way, we provide a novel operator-theoretic perspective to the event-triggering mechanism. The well known small gain theorem can be utilized to find a new condition under which the even-triggered multi-agent system reaches consensus.

The robust event-triggered consensus problem of multi-agent systems subject to additive node uncertainties and multiplicative network uncertainties in $\mathcal{RH}_\infty$ are considered in Sects. 7.2 and 7.3, respectively. In both cases, the robust event-triggered consensus problem can be reformulated as the robust stability problem of a high-dimensional system with two blocks of uncertainties, one of which describes the node or network uncertainties and the other corresponds to the linear operator generated by the event-triggered sampling mechanism. These two blocks of uncertainties are both linear finite-gain $\mathcal{L}_2$ stable operators in the frequency domain. By using the μ analysis tool, we obtain quantitative relationships among the parameters of the controllers, the Laplacian matrix of the network, and the robustness against aperiodic event triggering and frequency-domain uncertainties for both the cases of additive node uncertainties and multiplicative network uncertainties.

In Sect. 7.4, the event-triggered dynamic average consensus problem is further investigated, where the sampling errors are shown to be images of nonlinear finite-gain $\mathcal{L}_2$ stable operators. The robust performance of the proposed dynamic average consensus algorithm is analyzed, which indicates that the robustness and the performance are negatively related to the eigenratio of the Laplacian matrix. Simulation examples are finally provided in Sect. 7.5 to verify the obtained results.

7.1 An Operator-Theoretic Perspective to the Event-Triggering Mechanism

For simplicity, in this chapter we consider a network consisting of N single-input-single-output agents with scalar states. The dynamics of each agent can be described by a single integrator:

$$\dot{x}_i = u_i, \ i = 1, \cdots, N, \tag{7.1}$$

with x_i as the state variable of agent i and u_i the control input. It is assumed that the communication among the agents is depicted by a graph $\mathcal{G}$.

Assumption 7.1 The graph $\mathcal{G}$ is undirected and connected.

Under the event-triggering mechanism, agent i updates the estimate of its state $\hat{x}_i$ to the real value of x_i and sends it to its neighbors only at the triggering instants, between which the estimate is calculated locally by itself and its neighbors. We set the initial time t_0 as the first triggering instant of each agent and define a triggering function as follows:

$$f_i = e_i^2 - \alpha \sum_{j=1}^{N} a_{ij}(\hat{x}_i - \hat{x}_j)^2 - \mu e^{-\nu t},\ i = 1, \cdots, N, \tag{7.2}$$

where α, μ, ν are positive constants and e_i denotes the gap between the estimate $\hat{x}_i$ and the real state x_i, i.e., $e_i = \hat{x}_i - x_i$. The next event will happen whenever the triggering condition $f_i \geq 0$ is satisfied, i.e., $t_{k+1}^i \triangleq \inf\{t \mid t > t_k^i, f_i \geq 0\}$.

In this section, we consider the following distributed control law:

$$u_i = -\beta \sum_{j=1}^{N} a_{ij}(\hat{x}_i - \hat{x}_j),\ i = 1, \cdots, N. \tag{7.3}$$

During the time interval between two triggering instants, the estimate $\hat{x}_i$ used by all its neighbors is held to be a constant, i.e., $\hat{x}_i(t) = x_i(t_k^i)$, $\forall t \in [t_k^i, t_{k+1}^i)$. The control objective is to ensure that the states of the agents reach consensus, i.e., $x_i - x_j \to 0$, $\forall i, j \in \{1, \cdots, N\}$ as $t \to \infty$.

The closed-loop system derived from (7.1) and (7.3) is

$$\dot{x}_i = -\beta \sum_{j=1}^{N} a_{ij}(\hat{x}_i - \hat{x}_j),\ i = 1, \cdots, N. \tag{7.4}$$

Define $z_i = \frac{1}{N} \sum_{j=1}^{N} (\hat{x}_i - \hat{x}_j)$. Then, it follows that

$$\begin{aligned} \dot{x}_i &= -\beta \sum_{j=1}^{N} a_{ij} \left[(x_i - x_j) + (e_i - e_j) \right], \\ z_i &= \frac{1}{N} \sum_{j=1}^{N} [(x_i - x_j) + (e_i - e_j)]. \end{aligned} \tag{7.5}$$

We can further rewrite (7.5) in a compact form as

$$\begin{aligned} \dot{x} &= -\beta L x - \beta L e, \\ z &= M x + M e, \end{aligned} \tag{7.6}$$

where $M = I_N - \frac{1}{N} \mathbf{1_N} \mathbf{1_N^T}$ and L is the Laplacian matrix of the graph $\mathcal{G}$.

It is clear from the triggering function (7.2) that e_i is reset to be zero at each triggering instant and increases from 0 to some positive value during two triggering instants, and then drops again to zero at the next triggering instant. The square of the sampling error e_i is bounded from above by a quadratic form of $z(t)$ and an exponential decaying term at any time instant. While this bound relationship is described by the time-domain terminology, we discover a frequency-domain relationship between $e(s)$ and $z(s)$, as will be unveiled in the next theorem.

Theorem 7.1 *For the closed-loop network* (7.6) *with the triggering function* (7.2), *it follows that* $e(s) = \Delta(z(s))$, *where* $\Delta(\cdot) : \mathcal{L}_2 \mapsto \mathcal{L}_2$ *is a linear operator in the frequency domain with* $\|\Delta(z)\|_2 \leq \sqrt{2\alpha\lambda_N}\|z\|_2 + \sqrt{\frac{N\mu}{\nu}}$, *i.e.,* $\|\Delta\|_\infty \leq \sqrt{2\alpha\lambda_N}$, *where* λ_N *denotes the largest eigenvalue of* L.

Proof From (7.4) and the zero-order hold mechanism of $\hat{x}_i$, it is not difficult to obtain that

$$\begin{aligned} e_i(t) = \hat{x}_i(t) - x_i(t) &= \beta \sum_{j=1}^{N} a_{ij} \int_{t_k^i}^{t} (\hat{x}_i(\tau) - \hat{x}_j(\tau)) d\tau \\ &= \beta L_i^T \int_{t_k^i}^{t} \hat{x}(\tau) d\tau = \beta L_i^T M \int_{t_k^i}^{t} \hat{x}(\tau) d\tau \\ &= \beta L_i^T \int_{t_k^i}^{t} 1(t-\tau) z(\tau) d\tau, \end{aligned}$$

where L_i^T denotes the i-th row of the Laplacian matrix and $1(t) = 1$ when $t > 0$ and $1(t) = 0$ when $t < 0$. Therefore, $e_i(t) = \phi_i(z)$ where $\phi_i(\cdot) : \mathcal{L}_2 \mapsto \mathcal{L}_2$ is a linear operator in the time domain. In light of Lemma 2.7, we have $e_i(s) = \Delta_i(z(s))$, where $\Delta_i(\cdot) : \mathcal{L}_2 \mapsto \mathcal{L}_2$ denotes a linear operator in the frequency domain and thus we have $e(s) = \Delta(z(s))$.

Note that

$$\|e\|_2^2 = \sum_{i=1}^{N} \|e_i\|_2^2 = \sum_{i=1}^{N} \int_0^\infty e_i^*(t) e_i(t) dt.$$

According to the triggering condition (7.2), for any time instant t, we have

$$e_i^2 \leq \alpha \sum_{j=1}^{N} a_{ij} (\hat{x}_i - \hat{x}_j)^2 + \mu e^{-\nu t}.$$

Therefore,

$$
\begin{aligned}
\|e\|_2^2 &\le \sum_{i=1}^{N} \int_0^{\infty} \left(\alpha \sum_{j=1}^{N} a_{ij}(\hat{x}_i - \hat{x}_j)^2 + \mu e^{-\nu t} \right) dt \\
&= \int_0^{\infty} \left(\alpha \sum_{i=1}^{N} \sum_{j=1}^{N} a_{ij}(\hat{x}_i - \hat{x}_j)^2 + N\mu e^{-\nu t} \right) dt \\
&= \int_0^{\infty} \left(2\alpha z^T L z + N\mu e^{-\nu t} \right) dt \\
&\le 2\alpha\lambda_N \|z\|_2^2 + \frac{N\mu}{\nu} \\
&\le \left(\sqrt{2\alpha\lambda_N}\|z\|_2 + \sqrt{\frac{N\mu}{\nu}} \right)^2 .
\end{aligned}
$$

This completes the proof. □

Remark 7.1 The importance of this theorem lies in the following aspects. First, it illustrates that the sampling errors can be seen as the images of linear finite-gain operators in the frequency domain acting on the consensus errors of the sampled states. This paves the way to examine the robustness of the event-triggered algorithm to frequency-domain uncertainties. These operators, generally speaking, are not rational transfer matrices in $\mathcal{RH}_\infty$ as in the classic robust control. Nevertheless, these operators are finite-gain $\mathcal{L}_2$ stable. Second, it uncovers a quantitative relationship among the sampling parameter α, the largest eigenvalue of the Laplacian matrix, and the $\mathcal{L}_2$ gain of the operators which quantifies the effect of aperiodic event-triggering. More specifically, the larger λ_N and α are, the larger the $\mathcal{L}_2$ gain of the operators will be.

Remark 7.2 The event-triggered consensus algorithm has an equivalent block structure shown in Fig. 7.1. Note that the effect of the event-triggered mechanism is actually equivalent to introducing the virtual additional feedback loop in the dotted green block. The transfer function from x to $\hat{x}$ is $R = (I - \Delta M)^{-1}$, i.e., $\hat{x} = (I - \Delta M)^{-1}x$. When there is no event triggering, the operator Δ is equal to zero and thus $R = I$. The event-triggered consensus algorithm (7.4) then reduces to the classical one with continuous communication as in [76].

In the following theorem, the robust control tools are utilized to give a condition under which the network system reaches consensus under the event-triggered protocol (7.3).

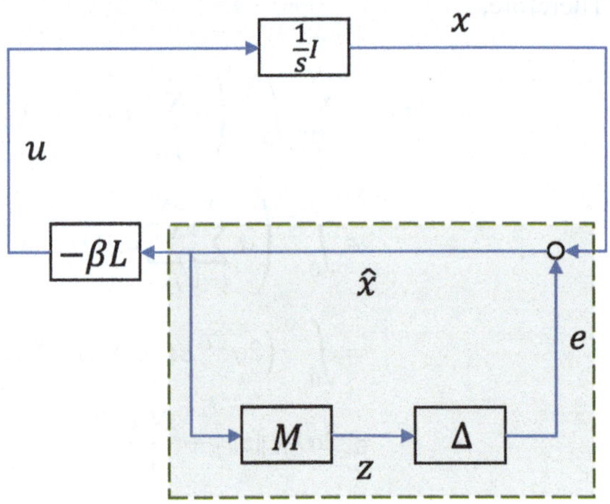

Fig. 7.1 An operator -theoretic reformulation of the distributed event-triggered consensus algorithm

Theorem 7.2 *The network system* (7.1) *reaches consensus under the event-triggered control law* (7.3)*, if the parameter* α *in the triggering function* f_i *is selected to satisfy* $2\alpha\lambda_N < 1$. *Moreover, the closed-loop system does not exhibit the Zeno behavior as long as* $\mu > 0$ *and* $\nu > 0$.

Proof Letting U be the unitary matrix such that $U^T LU = \Lambda = \text{diag}\{0, \lambda_2, \cdots, \lambda_N\} = \text{diag}\{0, \bar{\Lambda}\}$, and denoting $\tilde{x} = U^T x$, $\tilde{z} = U^T z$, $\tilde{e} = U^T e$, it is easy to find that consensus is reached if and only if $\tilde{x}_i$ is asymptotically stable $\forall i = 2, \cdots, N$. Setting $\bar{x} = \tilde{x}_{2:N}$, $\bar{z} = \tilde{z}_{2:N}$, $\bar{e} = \tilde{e}_{2:N}$, where $(\cdot)_{2:N}$ denotes the subvector that takes the second to the N-th elements of the original vector. We can then derive from (7.5) that

$$\begin{aligned} \dot{\bar{x}} &= -\beta\bar{\Lambda}\bar{x} - \beta\bar{\Lambda}\bar{e}, \\ \bar{z} &= \bar{x} + \bar{e}. \end{aligned} \tag{7.7}$$

Note also that U can be written as $\begin{bmatrix} \frac{1_N}{\sqrt{N}} & Y \end{bmatrix}$ with $Y^T Y = I_{N-1}$ and $YY^T = M$. It is easy to see that

$$\tilde{e} = U^T \Delta(UU^T z) = \begin{bmatrix} \frac{1^T}{\sqrt{N}} \\ Y^T \end{bmatrix} \Delta(\frac{1}{\sqrt{N}}\tilde{z}_1 + Y\bar{z}) = \begin{bmatrix} \frac{1^T}{\sqrt{N}} \\ Y^T \end{bmatrix} \Delta(Y\bar{z}),$$

where $\tilde{z}_1 \equiv 0$ is used to get the last equality. Then, we have

$$\bar{e} = Y^T \Delta(Y\bar{z}) = \bar{\Delta}(\bar{z}), \tag{7.8}$$

where $\bar{\Delta}$ is a linear operator.

It is not difficult to verify that

$$\begin{aligned}\|\bar{e}\|_2^2 &= \int_0^\infty \Delta^*(Y\bar{z})M\Delta(Y\bar{z})dt \\ &\leq \|\Delta(Y\bar{z})\|_2^2 \leq \left(\sqrt{2\alpha\Lambda_N}\|Y\bar{z}\|_2 + \sqrt{\frac{N\mu}{\nu}}\right)^2 \\ &= \left(\sqrt{2\alpha\lambda_N}\|\bar{z}\|_2 + \sqrt{\frac{N\mu}{\nu}}\right)^2.\end{aligned}$$

Therefore, we have

$$\|\bar{\Delta}(\bar{z})\|_2 \leq \sqrt{2\alpha\lambda_N}\|\bar{z}\|_2 + \sqrt{\frac{N\mu}{\nu}},$$

implying that $\|\bar{\Delta}\|_\infty \leq \sqrt{2\alpha\lambda_N}$.

In light of Lemma 2.8, the system is internally stable if $\|T_{\bar{e}\bar{z}}\|_\infty\|\bar{\Delta}\|_\infty < 1$, where $T_{\bar{e}\bar{z}}$ denotes the transfer matrix from $\bar{e}$ to $\bar{z}$, given by

$$T_{\bar{e}\bar{z}} = -(sI + \beta\bar{\Lambda})^{-1}\beta\bar{\Lambda} + I = (sI + \beta\bar{\Lambda})^{-1}sI.$$

Note that

$$\|T_{\bar{e}\bar{z}}\|_\infty = \max_{i=2,\cdots,N}\left\|\frac{s}{s+\beta\lambda_i}\right\|_\infty = 1.$$

Therefore, the system (7.7) is internally stable, if $\|\bar{\Delta}\|_\infty < 1$, which is satisfied if $2\alpha\lambda_N < 1$.

Next, we exclude the Zeno behavior. Notice that during each time interval between any two consecutive triggering instants, i.e., $[t_k^i, t_{k+1}^i)$, $\dot{e}_i = -\dot{x}_i = \beta\sum_{j=1}^N a_{ij}(\hat{x}_i - \hat{x}_j)$ is bounded by a positive real number, say H. Suppose that there exists the Zeno behavior. Then there exists an agent i, such that $\lim_{k\to\infty} t_k^i = T < \infty$. Thus, for a small positive number $\delta < \frac{\sqrt{\mu}e^{-\frac{\nu T}{2}}}{H}$, there exists a positive integer K such that for $\forall k \geq K$, $t_k^i \in (T-\delta, T]$. Notice that at the triggering instant t_K^i, $e_i = 0$. And the next triggering time is the first time when e_i^2 reaches $\alpha\sum_{j=1}^N a_{ij}(\hat{x}_i - \hat{x}_j)^2 + \mu e^{-\nu t}$. Then, there must exist some time instant T_0 when $e_i^2(T_0) = \mu e^{-\nu T_0}$. Since

$$e_i^2(T_0) = (x_i(t_K^i) - x_i(T_0))^2 = \left(\int_{t_K^i}^{T_0}\beta\sum_{j=1}^N(\hat{x}_i - \hat{x}_j)dt\right)^2,$$

we have

$$e_i^2(T_0) \leq H^2(T_0 - t_K^i)^2.$$

On the other hand, $\mu e^{-\nu T_0} \geq \mu e^{-\nu T}$. Thus, we have

$$H^2(T_0 - t_K^i)^2 \geq \mu e^{-\nu T},$$

which implies

$$t_{K+1}^i - t_K^i \geq T_0 - t_K^i \geq \frac{\sqrt{\mu} e^{-\frac{\nu T}{2}}}{H}.$$

Since $\delta < \frac{\sqrt{\mu} e^{-\frac{\nu T}{2}}}{H}$, it follows that the $(K+1)$-th triggering instant $t_{K+1} > t_K^i + \delta > T$, which leads to a contradiction. □

Remark 7.3 This theorem unveils some essential requirements that the gain of the operator should not be too large to achieve consensus under the event-triggered protocol (7.3). The operator gain characterizes the extent of the sampling error introduced by the aperiodic event triggering. The larger the operator gain is, the larger the $\mathcal{L}_2$ norm of the sampling errors will be. From the quantitative relationship shown in this theorem, if λ_N is larger, then it is more reliable to trigger more frequently in the sense that α should be smaller.

Remark 7.4 Different from most of the previous works, e.g., [3, 21, 28, 73, 105], where the time-domain Lyapunov stability analysis is used, in this section an operator is constructed to characterize the relationship between the sampling error e and the variable z and the small gain theorem is utilized to get the consensus condition. Interestingly, since $\lambda_N \leq 2d_i$, the consensus condition in Theorem 7.2 is less conservative than that of [73]. More importantly, this method provides a feasible way to handle the robustness of event-triggered control when the network systems are subject to frequency-domain uncertainties, as will be shown in the next sections.

7.2 Robustness Against Additive Node Uncertainties

In the previous section, we analyze the event-triggered consensus problem of the integrator network without uncertainties in a frequency-domain approach. In this section, we consider multi-agent systems under additive node uncertainties.

The agent dynamics in this case are described by

$$\begin{aligned} \dot{x}_i &= u_i, \\ y_i &= x_i + d_i, \\ d_i(s) &= \Delta_i^a(u_i(s)),\ i = 1, \cdots, N, \end{aligned} \tag{7.9}$$

where Δ_i^a is a linear finite-gain $\mathcal{L}_2$ stable operator with operator gain $\|\Delta_i^a\|_\infty \leq \eta$. Note that this definition includes the transfer matrices in $\mathcal{RH}_\infty$ as special cases.

We consider the following distributed output feedback protocol:

$$u_i = -\beta \sum_{j=1}^{N} a_{ij}(\hat{y}_i - \hat{y}_j),\ i = 1, \cdots, N, \tag{7.10}$$

where $\hat{y}_i$ denotes the estimate of the output y_i, which is updated to the real value $y_i(t_k^i)$ and broadcasted to all its neighbors at the k-th triggering instant of the agent i, i.e., t_k^i, and keeps constant during two triggering instants. Define $\epsilon_i = \hat{y}_i - y_i$. Moreover, the triggering function of agent i is set to be

$$f_i = \epsilon_i^2 - \alpha \sum_{j=1}^{N} a_{ij}(\hat{y}_i - \hat{y}_j)^2 - \mu e^{-\nu t}. \tag{7.11}$$

According to (7.9) and (7.10), we have

$$\dot{x}_i = -\beta \sum_{j=1}^{N} a_{ij} \left[(x_i - x_j) + (d_i - d_j) + (\epsilon_i - \epsilon_j) \right]. \tag{7.12}$$

Denote $w_i = \frac{1}{N} \sum_{j=1}^{N} (\hat{y}_i - \hat{y}_j)$. We can obtain the closed-loop network dynamics as follows:

$$\begin{aligned} \dot{x} &= -\beta L x - \beta L d - \beta L \epsilon, \\ u &= -\beta L x - \beta L d - \beta L \epsilon, \\ w &= M x + M d + M \epsilon. \end{aligned} \tag{7.13}$$

Theorem 7.3 *For the network* (7.9) *under the control law* (7.10) *with the triggering function* (7.11), *it follows that* $\epsilon(s) = \Delta^b(w(s))$, *where* $\Delta^b(\cdot) : \mathcal{L}_2 \mapsto \mathcal{L}_2$ *is a linear operator with the operator gain* $\|\Delta^b\|_\infty \leq \sqrt{2\alpha\lambda_N}$.

Proof From the definition of ϵ_i, we know that $\epsilon_i(t) = \hat{y}_i - y_i = y(t_k^i) - y_i(t) = x_i(t_k^i) + d_i(t_k^i) - x_i(t) - d_i(t)$. According to (7.9) and (7.10), when $t \in [t_k^i, t_{k+1}^i)$,

$$
\begin{aligned}
x_i(t) - x_i(t_{k_i}^i) &= -\beta \int_{t_{k_i}^i}^{t} \sum_{j=1}^{N} a_{ij}(\hat{y}_i(\tau) - \hat{y}_j(\tau))d\tau \\
&= -\beta \int_{t_k^i}^{t} L_i^T \hat{y}(\tau)d\tau \\
&= -\beta L_i^T \int_{t_k^i}^{t} w(\tau)d\tau.
\end{aligned}
$$

Thus, we have $x_i(t) - x_i(t_{k_i}^i) = \psi_i(w(t))$, where $\psi_i(\cdot) : \mathcal{L}_2 \mapsto \mathcal{L}_2$ is a linear operator. Since $d_i(s) = \Delta_i^a(u_i(s))$, by letting $d_i(t) = \tilde{\Delta}_i^a(u_i(t))$ and $d(t_k^i) = \bar{\Delta}_i^a(u_i(t))$, we obtain that $\tilde{\Delta}_i^a(\cdot)$ and $\bar{\Delta}_i^a(\cdot)$ are linear operators in the time domain. It is then not difficult to find that $d_i(t_{k_i}^i) - d_i(t)$ is a linear operator of $u_i(t)$.

Denote $d_i(t_k^i) - d_i(t) = \phi_i(u_i(t))$ and note that $u_i(t) = -\beta \sum_{j=1}^{N} a_{ij}(\hat{y}_i(t) - \hat{y}_j(t)) = -\beta L_i^T \hat{y}(t) = -\beta L_i^T w(t)$. Thus,

$$
d_i(t_k^i) - d_i(t) = \phi_i(-\beta L_i^T w(t)) = \phi_i'(w(t)),
$$

and it is evident that $\phi_i'(\cdot) : \mathcal{L}_2 \mapsto \mathcal{L}_2$ is a linear operator. Therefore, $\epsilon_i = -\psi_i(w(t)) + \phi_i'(w(t)) = \Psi_i(w(t))$, where $\Psi_i'(\cdot) : \mathcal{L}_2 \mapsto \mathcal{L}_2$ is a linear operator. According to Lemma 2.7, $\epsilon_i(s) = \Delta_i^b(w(s))$, where $\Delta_i^b(\cdot) : \mathcal{L}_2 \mapsto \mathcal{L}_2$ is a linear operator in the frequency domain. Moreover, $\epsilon = \Delta^b(w(s))$, where $\Delta : \mathcal{L}_2 \mapsto \mathcal{L}_2$ is a linear operator. The operator gain $\|\Delta^b\|_\infty$ can be similarly determined as in Theorem 7.1 and is omitted here for brevity.

□

Remark 7.5 It is worth noting that two blocks of operators Δ^a and Δ^b appearing in Fig. 7.2 are essentially different. The first block of operator represents the uncertainties of the agent dynamics in $\mathcal{RH}_\infty$, possibly caused by model uncertainties or unmodeled dynamics. The second block of operator Δ^b is caused by the event-triggered sampling mechanism and in general does not belong to $\mathcal{RH}_\infty$. Quite interestingly, these two blocks of operators generated by totally different mechanisms can be unified in an operator-theoretic framework, since the two blocks of uncertainties are both linear finite-gain $\mathcal{L}_2$ stable operators in the frequency domain.

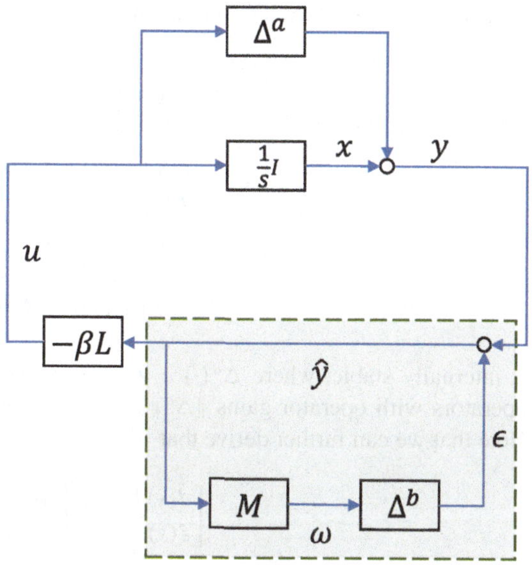

Fig. 7.2 Event-triggered network system with additive dynamic uncertainties

As will shown in the next theorem, the two blocks of operators in Fig. 7.2 cannot be too large in order to guarantee the robust consensus of the network.

Theorem 7.4 *Let* $\gamma = \max\{\eta, \sqrt{2\alpha\lambda_N}\}$. *The uncertain network* (7.9) *reaches robust consensus under the event-triggered control law* (7.10) *with the triggering function* (7.11), *if the positive constants* β, α, μ, *and* v *satisfy that* $(\beta\lambda_N + 1)\gamma < 1$. *Moreover, the closed-loop system does not exhibit the Zeno behavior.*

Proof Notice that the interconnecting system can be rewritten into the following form:

$$
\begin{aligned}
\dot{x} &= -\beta L x + \begin{bmatrix} -\beta L & -\beta L \end{bmatrix} \begin{bmatrix} d \\ \epsilon \end{bmatrix}, \\
\begin{bmatrix} u \\ w \end{bmatrix} &= \begin{bmatrix} -\beta L \\ M \end{bmatrix} x + \begin{bmatrix} -\beta L & -\beta L \\ M & M \end{bmatrix} \begin{bmatrix} d \\ \epsilon \end{bmatrix}, \\
d(s) &= \Delta^a(u(s)), \\
\epsilon(s) &= \Delta^b(w(s)).
\end{aligned}
\tag{7.14}
$$

Here $\Delta^a(\cdot) = \operatorname{diag}\{\Delta_1^a, \cdots, \Delta_N^a\}$ is also a linear operator with operator gain $\|\Delta^a\|_\infty \leq \eta$. Similarly as in Theorem 7.2, denoting $\tilde{\epsilon} = U^T \epsilon$, $\tilde{w} = U^T w$, $\bar{\epsilon} = \tilde{\epsilon}_{2:N}$, and $\bar{w} = \tilde{w}_{2:N}$, we can get that the system (7.14) reaches robust consensus if and only if the following system interconnection:

$$
\begin{aligned}
\dot{\bar{x}} &= -\beta\bar{\Lambda}\bar{x} + \begin{bmatrix} -\beta\bar{\Lambda} & -\beta\bar{\Lambda} \end{bmatrix} \begin{bmatrix} \bar{d} \\ \bar{\epsilon} \end{bmatrix}, \\
\begin{bmatrix} \bar{u} \\ \bar{w} \end{bmatrix} &= \begin{bmatrix} -\beta\bar{\Lambda} \\ I \end{bmatrix} \bar{x} + \begin{bmatrix} -\beta\bar{\Lambda} & -\beta\bar{\Lambda} \\ I & I \end{bmatrix} \begin{bmatrix} \bar{d} \\ \bar{\epsilon} \end{bmatrix}, \\
\bar{d}(s) &= Y^T \Delta^a (Y\bar{u}(s)) = \bar{\Delta}^a(\bar{u}(s)), \\
\bar{\epsilon}(s) &= Y^T \Delta^b (Y\bar{w}(s)) = \bar{\Delta}^b(\bar{w}(s)),
\end{aligned} \tag{7.15}
$$

is internally stable, where $\bar{\Delta}^a(\cdot) : \mathcal{L}_2 \mapsto \mathcal{L}_2$ and $\bar{\Delta}^b(\cdot) : \mathcal{L}_2 \mapsto \mathcal{L}_2$ are linear operators with operator gains $\|\bar{\Delta}^a\|_\infty \le \eta$ and $\|\bar{\Delta}^b\|_\infty \le \sqrt{2\alpha\lambda_N}$, respectively. Note that we can further derive that

$$
\begin{bmatrix} \bar{d}(s) \\ \bar{\epsilon}(s) \end{bmatrix} = \mathbf{\Delta} \begin{bmatrix} \bar{u}(s) \\ \bar{w}(s) \end{bmatrix}, \tag{7.16}
$$

where $\mathbf{\Delta}$ is a linear block diagonal operator defined as

$$
\mathbf{\Delta} = \begin{bmatrix} \bar{\Delta}^a(\cdot) & 0(\cdot) \\ 0(\cdot) & \bar{\Delta}^b(\cdot) \end{bmatrix},
$$

and $\|\mathbf{\Delta}\|_\infty \le \gamma$.

In light of Lemma 2.9, the system (7.15) is internal stable, if $\mu_{\mathbf{\Delta}}(G(j\omega)) < \frac{1}{\gamma}$, where G is the transfer matrix from $\begin{bmatrix} \bar{d}^T & \bar{\epsilon}^T \end{bmatrix}^T$ to $\begin{bmatrix} \bar{u}^T & \bar{w}^T \end{bmatrix}^T$. From (7.15), it is easy to derive that

$$
G(s) = \begin{bmatrix} -\beta s\bar{\Lambda}(sI + \beta\bar{\Lambda})^{-1} & -\beta s\bar{\Lambda}(sI + \beta\bar{\Lambda})^{-1} \\ s(sI + \beta\bar{\Lambda})^{-1} & s(sI + \beta\bar{\Lambda})^{-1} \end{bmatrix} \triangleq \begin{bmatrix} G_{11}(s) & G_{12}(s) \\ G_{21}(s) & G_{22}(s) \end{bmatrix},
$$

where

$$
\begin{aligned}
\|G_{11}(j\omega)\|^2 = \|G_{12}(j\omega)\|^2 &= \|j\omega\beta\bar{\Lambda}(j\omega I + \beta\bar{\Lambda})^{-1}\|^2 \\
&= \max_{i=2,\cdots,N} \frac{\beta^2\lambda_i^2\omega^2}{\omega^2 + \beta^2\lambda_i^2} = \frac{\beta^2\lambda_N^2\omega^2}{\omega^2 + \beta^2\lambda_N^2},
\end{aligned}
$$

and

$$
\|G_{22}(j\omega)\|^2 = \|G_{21}(j\omega)\|^2 = \|j\omega(j\omega I + \beta\bar{\Lambda})^{-1}\|^2 = \frac{\omega^2}{\omega^2 + \beta^2\lambda_2^2}.
$$

According to Lemma 2.10, the system interconnection (7.15) is internally stable, if $\frac{\beta\lambda_N|\omega|}{\sqrt{\omega^2+\beta^2\lambda_N^2}} + \frac{|\omega|}{\sqrt{\omega^2+\beta^2\lambda_2^2}} < \frac{1}{\gamma}$, which holds if $\beta\lambda_N + 1 < \frac{1}{\gamma}$. Next, it remains to rule out the Zeno behavior, which is quite similar to the counterpart of Theorem 7.2. The details are omitted here for conciseness. □

Remark 7.6 Though the time-domain sampling mechanism and the frequency-domain uncertainties seem incompatible, they can indeed be tackled in a unified way under our operator-theoretic framework. This theorem quantitatively characterizes the relationship among Laplacian matrix, the output feedback gain and the robustness of the event-triggered consensus algorithm against event triggering and frequency-domain uncertainties. From this theorem, it is not difficult to find that the smaller λ_N and the parameter β are, the larger γ can be, meaning that the more robust the network is against frequency-domain uncertainties and sparser triggering.

7.3 Robustness Against Multiplicative Edge Uncertainties

In the previous sections, it is assumed that the agents interact via a fixed and known graph $\mathcal{G}$. In practice, however, the topology graph may be subject to various kinds of perturbations, which will render the network graph uncertain and time-varying [52]. In this section, we consider a network graph with uncertain communication strengths. Specifically, each element a_{ij} in the adjacency matrix cannot be known exactly but is rather perturbed to a bounded region that can be expressed in the form of $a_{ij}(1 + \Delta_{ij}^c(\cdot))$, where $\Delta_{ij}^c(\cdot) : \mathcal{L}_2 \mapsto \mathcal{L}_2$ is a linear finite-gain operator in the time domain with operator norm $\|\Delta_{ij}^c\|_\infty \leq \delta < 1$.

In this case, the distributed control law (7.3) then becomes

$$\begin{aligned} u_i &= -\beta \sum_{j=1}^{N} a_{ij}(\hat{x}_i - \hat{x}_j) - \beta \sum_{j=1}^{N} a_{ij}\Delta_{ij}^c(\hat{x}_i - \hat{x}_j) \\ &= -\beta \sum_{j=1}^{N} a_{ij}(\hat{x}_i - \hat{x}_j) - \beta \sum_{j=1}^{N} \left(a_{ij}^{\frac{1}{2}}\Delta_{ij}^c(a_{ij}^{\frac{1}{2}}\hat{x}_i) - a_{ij}^{\frac{1}{2}}\Delta_{ij}^c(a_{ij}^{\frac{1}{2}}\hat{x}_j) \right) \\ &= -\beta L_i^T \hat{x} - \beta D_i W^{\frac{1}{2}} \Delta^c(W^{\frac{1}{2}} D^T \hat{x}), \end{aligned} \tag{7.17}$$

where D_i is the i-th row of the incidence matrix of the graph $\mathcal{G}$ and $\Delta^c(\cdot)$ is the block diagonal linear finite-gain operator with $\Delta_{ij}^c(\cdot)$ as its diagonal elements. It is easy to see that $\|\Delta^c\|_\infty \leq \delta$. Letting $v = \Delta^c(W^{\frac{1}{2}} D^T \hat{x})$ and $w = W^{\frac{1}{2}} D^T \hat{x}$, we have $v = \Delta^c(w)$. According to Lemma 2.7, $v(s) = \bar{\Delta}^c(w(s))$.

Define the sampling error $e_i = \hat{x}_i - x_i$. The triggering function of agent i is given by

$$f_i = e_i^2 - \alpha \sum_{j=1}^{N} a_{ij}(\hat{x}_i - \hat{x}_j)^2 - \mu e^{-\nu t}. \tag{7.18}$$

Recalling (7.1), the system interconnection can be rewritten in the following compact form:

$$\begin{aligned} \dot{x} &= -\beta L x - \beta L e - \beta D W^{\frac{1}{2}} v, \\ w &= W^{\frac{1}{2}} D^T x + W^{\frac{1}{2}} D^T e, \\ z &= Mx + Me, \\ v(s) &= \bar{\Delta}^c(w(s)). \end{aligned} \tag{7.19}$$

Similarly, we have

$$\begin{aligned} e_i(t) &= \beta \int_{t_k^i}^{t} L_i^T \hat{x}(\tau) d\tau + \beta \int_{t_k^i}^{t} D_i W^{\frac{1}{2}} \Delta^c(W^{\frac{1}{2}} D^T \hat{x}(\tau)) d\tau \\ &= \beta \int_{t_k^i}^{t} D_i W z(\tau) d\tau + \beta \int_{t_k^i}^{t} D_i W^{\frac{1}{2}} \Delta^c(W^{\frac{1}{2}} z(\tau)) d\tau \\ &\quad - \Delta^d(z(t)), \end{aligned}$$

where $\Delta^d(\cdot) : \mathcal{L}_2 \mapsto \mathcal{L}_2$ is a linear operator in the time domain, following the proof of Theorem 7.1.

In virtue of Lemma 2.7, $e_i(s) = \bar{\Delta}^d(z(s))$, where $\bar{\Delta}^d(\cdot) : \mathcal{L}_2 \mapsto \mathcal{L}_2$ is a linear operator in the frequency domain. Then, in light of Theorem 7.1, it is easy to see that $\|\bar{\Delta}\|_\infty^d \leq \sqrt{2\alpha\lambda_N}$. Moreover,

$$\begin{aligned} \|\bar{\Delta}^c\|_\infty^2 &= \sup_{\|w\|_2=1} \frac{1}{2\pi} \int_{-\infty}^{\infty} v^*(j\omega) v(j\omega) d\omega \\ &= \sup_{\|w\|_2=1} \int_0^{\infty} v^*(t) v(t) dt = \|\Delta^c\|_\infty^2 \leq \delta^2. \end{aligned}$$

We have $\|\bar{\Delta}^c\|_\infty \leq \delta$.

Theorem 7.5 *The network* (7.1) *reaches consensus under the event-triggered control law* (7.17) *with the triggering function* (7.18)*, if* $\gamma < \sqrt{\frac{\lambda_2^2}{\lambda_2^2 + \lambda_N^2}}$ *where* $\gamma = \max\{\delta, \sqrt{2\alpha\lambda_N}\}$. *Moreover, the closed-loop system does not exhibit the Zeno behavior.*

Proof Letting $\xi = Mx$, the system interconnection can be written in the following form:

$$\begin{aligned}
\dot{\xi} &= -\beta L\xi + \begin{bmatrix} -\beta D W^{\frac{1}{2}} & -\beta L \end{bmatrix} \begin{bmatrix} v \\ e \end{bmatrix}, \\
\begin{bmatrix} w \\ z \end{bmatrix} &= \begin{bmatrix} W^{\frac{1}{2}} D^T \\ M \end{bmatrix} \xi + \begin{bmatrix} 0 & W^{\frac{1}{2}} D^T \\ 0 & M \end{bmatrix} \begin{bmatrix} v \\ e \end{bmatrix}, \\
\begin{bmatrix} v(s) \\ e(s) \end{bmatrix} &= \begin{bmatrix} \bar{\Delta}^c & \\ & \bar{\Delta}^d \end{bmatrix} \begin{bmatrix} w(s) \\ z(s) \end{bmatrix}.
\end{aligned} \tag{7.20}$$

Note that the state x of the system reaches consensus if and only if ξ is asymptotically stable, i.e., the system interconnection (7.20) is internally stable. According to Lemma 2.9, it is enough to let $\mu_{\mathbf{\Delta}}(G(j\omega)) < \frac{1}{\gamma}$, where G is the transfer matrix from $\begin{bmatrix} v^T & e^T \end{bmatrix}^T$ to $\begin{bmatrix} w^T & z^T \end{bmatrix}^T$ and

$$\begin{aligned}
G &\triangleq \begin{bmatrix} G_{11}(s) & G_{12}(s) \\ G_{21}(s) & G_{22}(s) \end{bmatrix} \\
&= \begin{bmatrix} -\beta W^{\frac{1}{2}} D^T (sI + \beta L)^{-1} D W^{\frac{1}{2}} & s W^{\frac{1}{2}} D^T (sI + \beta L)^{-1} \\ -\beta M (sI + \beta L)^{-1} D W^{\frac{1}{2}} & s M (sI + \beta L)^{-1} \end{bmatrix}.
\end{aligned}$$

Note that

$$\begin{aligned}
\|G_{11}(j\omega)\|^2 &= \rho(\beta^2 W^{\frac{1}{2}} D^T (-j\omega I + \beta L)^{-1} L (j\omega I + \beta L)^{-1} D W^{\frac{1}{2}}) \\
&= \rho(\beta^2 (-j\omega I + \beta \Lambda)^{-1} \Lambda (j\omega I + \beta \Lambda)^{-1} \Lambda) \\
&= \frac{\beta^2 \lambda_N^2}{\omega^2 + \beta^2 \lambda_N^2} \le \frac{\beta^2 \lambda_N^2}{\omega^2 + \beta^2 \lambda_2^2}.
\end{aligned}$$

Similarly,

$$\|G_{12}(j\omega)\|^2 \le \frac{\omega^2 \lambda_N}{\omega^2 + \beta^2 \lambda_2^2},$$

$$\|G_{21}(j\omega)\|^2 = \max_{i=2,\cdots,N} \frac{\beta^2 \lambda_i}{\omega^2 + \beta^2 \lambda_i^2} \le \frac{\beta^2 \lambda_N}{\omega^2 + \beta^2 \lambda_2^2},$$

and

$$\|G_{22}(j\omega)\|^2 = \frac{\omega^2}{\omega^2 + \beta^2 \lambda_2^2}.$$

According to Lemma 2.10, the system interconnection is internally stable, if

$$\frac{(\beta\lambda_N + |\omega|)^2}{\beta^2\lambda_2^2 + \omega^2} < \frac{1}{\gamma^2}, \quad \forall\omega \in \mathbf{R}. \tag{7.21}$$

It is not difficult to find that a sufficient condition for (7.21) to hold is $\gamma < \sqrt{\frac{\lambda_2^2}{\lambda_2^2+\lambda_N^2}}$. The exclusion of the Zeno behavior is quite similar to the proof of Theorem 7.2. □

Remark 7.7 Different from the additive node uncertainty case considered in the previous section, the robustness of the event-triggered consensus algorithm against the network topology uncertainties is closely related to the eigenratio $\frac{\lambda_N}{\lambda_2}$ of the Laplacian matrix. When the eigenratio $\frac{\lambda_N}{\lambda_2}$ is smaller, the robust margin γ is larger in the sense that the network can tolerant larger network uncertainties and larger sampling errors.

Remark 7.8 Note that the uncertainties of the network topology can be either time-domain or frequency-domain, which are equivalent according to Lemma 2.7. Actually, it is not necessary for the uncertainties to be in $\mathcal{RH}_\infty$. As long as the uncertainties are finite-gain $\mathcal{L}_2$-stable operators, e.g., bounded communication delays or nonlinearities, the robust event-triggered consensus problem can be solved in our operator-theoretic framework.

7.4 Extensions to Dynamic Average Consensus

In Sects. 7.1 and 7.2, the control objective is to reach static average consensus, i.e., each state variable x_i converges to the average of the initial states of all agents. In this section, we generalize the results to the event-based dynamic average consensus problem. Different from the static average consensus problem, the dynamic average consensus problem aims to make the state of each agent converge to the average of the reference signals [49]. Till now, there have been a lot of dynamic average consensus algorithms proposed in the literatures for first-order network systems, e.g., [25, 49]. An event-triggered dynamic average consensus algorithm is provided in [48], where the agent model is assumed to be nominal and there are no uncertainties. Here, we consider the following network subject to additive dynamic uncertainties:

$$\begin{aligned} \dot{x}_i &= u_i + \dot{r}_i, \\ y_i &= x_i + d_i, \\ d_i(s) &= \Delta_i^a(u_i(s)), \ i = 1, \cdots, N, \end{aligned} \tag{7.22}$$

where r_i denote the reference signals and Δ_i^a is defined as in (7.9).

The objective of the robust dynamic average consensus problem considered in this section is to ensure that for the uncertain agents in (7.22), $x_i \to \frac{1}{N}\sum_{i=1}^{N} r_i, \forall i = 1, \cdots, N$ as $t \to \infty$.

Throughout this section, we suppose that the following assumption holds.

Assumption 7.2 The reference signals r_i are bounded.

To achieve the dynamic average consensus, the following event-triggered algorithm is utilized:

$$\begin{aligned}\dot{w}_i &= -\theta(w_i - y_i) - \beta \sum_{j=1}^{N} a_{ij}(\hat{w}_i - \hat{w}_j),\\ u_i &= -\beta \sum_{j=1}^{N} a_{ij}(\hat{w}_i - \hat{w}_j),\end{aligned} \tag{7.23}$$

where w_i is the augmented state variable.

To save the communication cost, the event-triggered scheme is used. At the k-th triggering instant t_k^i of the agent i, agent i updates its estimate of its own states $\hat{w}_i$ to $w(t_k^i)$ and its estimate of its own output $\hat{y}_i$ to $y(t_k^i)$ and broadcasts them to all its neighbors. During two triggering instants of the agent i, $\hat{w}_i$ is calculated in the following way:

$$\hat{w}_i(t) = e^{-\theta(t-t_k^i)} w(t_k^i) + \int_{t_k^i}^{t} e^{-\theta(t-\tau)} \theta \hat{y}_i(\tau) d\tau,$$

and

$$\hat{y}_i(t) = y(t_k^i).$$

The triggering function of the agent i is given by

$$f_i = \epsilon_i^2 - \alpha \sum_{j=1}^{N} a_{ij}(\hat{w}_i - \hat{w}_j)^2 - \mu e^{-\nu t}, \tag{7.24}$$

where $\epsilon_i = \hat{w}_i - w_i$ denotes the sampling error. Define $z_i = \frac{1}{N}\sum_{j=1}^{N}(\hat{w}_i - \hat{w}_j)$, $z = \begin{bmatrix} z_1 \cdots, z_N \end{bmatrix}^T$ and $\epsilon = \begin{bmatrix} \epsilon_1 \cdots, \epsilon_N \end{bmatrix}^T$. It then follows that $z = M\hat{w}$.

The following theorem describes the relationship between ϵ and z.

Theorem 7.6 *For the triggering function* (7.24), *it follows that* $\epsilon(s) = \Delta^e(z(s))$, *where* $\Delta^e(\cdot)$ *is a nonlinear finite-gain* $\mathcal{L}_2$ *stable operator with the operator norm* $\|\Delta^e\|_\infty \le \sqrt{2\alpha\lambda_N}$.

Proof It follows from (7.23) that for $t \in [t_k^i, t_{k+1}^i)$,

$$w_i(t) = e^{-\theta(t-t_k^i)}w(t_k^i) + \int_{t_k^i}^t e^{-\theta(t-\tau)}\theta y_i(\tau)d\tau$$
$$- \int_{t_k^i}^t e^{-\theta(t-\tau)}\beta \sum_{j=1}^N a_{ij}(\hat{w}_i(\tau) - \hat{w}_j(\tau))d\tau.$$

Then, we can obtain that

$$\begin{aligned}
\epsilon_i = \hat{w}_i - w_i = & \int_{t_k^i}^t e^{-\theta(t-\tau)}\beta \sum_{j=1}^N a_{ij}(\hat{w}_i(\tau) - \hat{w}_j(\tau))d\tau \\
& + \int_{t_k^i}^t e^{-\theta(t-\tau)}\theta(\hat{y}_i(\tau) - y_i(\tau))d\tau \\
= & \int_{t_k^i}^t e^{-\theta(t-\tau)}\theta((\hat{y}_i(\tau) - y_i(\tau)) - (\hat{r}_i(\tau) - r_i(\tau)))d\tau \\
& + \int_{t_k^i}^t e^{-\theta(t-\tau)}\theta(\hat{r}_i(\tau) - r_i(\tau))d\tau \\
& + \int_{t_k^i}^t e^{-\theta(t-\tau)}\beta \sum_{j=1}^N a_{ij}(\hat{w}_i(\tau) - \hat{w}_j(\tau))d\tau,
\end{aligned}$$

where $\hat{r}_i(t) = r_i(t_k^i)$, $\forall t \in [t_k^i, t_{k+1}^i)$.

On the other hand,

$$\begin{aligned}
\hat{y}_i - y_i &= x(t_k^i) + d(t_k^i) - (x(t) + d(t)) \\
&= d(t_k^i) - d(t) + \beta \int_{t_k^i}^t \sum_{j=1}^N a_{ij}(\hat{w}_i - \hat{w}_j)d\tau - r_i(t) + r_i(t_k^i).
\end{aligned}$$

Since $d_i(s) = \Delta_i^a(u_i(s))$, we have $d(t) = \tilde{\Delta}_i^a(u_i(t))$ and $d(t_k^i) = \bar{\Delta}_i^a(u_i(t))$, where $\tilde{\Delta}_i^a(\cdot)$ and $\bar{\Delta}_i^a(\cdot)$ are linear operators in the time domain. Notice that

$$\begin{aligned}
d_i(t_k^i) - d_i(t) &= \bar{\Delta}_i^a(u_i(t)) - \tilde{\Delta}_i^a(u_i(t)) = \Theta_i(u_i(t)) \\
&= \Theta_i(-\beta \sum_{j=1}^N a_{ij}(\hat{w}_i - \hat{w}_j)) = \Xi_i(z(t)),
\end{aligned}$$

where Θ, $\Xi(\cdot) : \mathcal{L}_2 \mapsto \mathcal{L}_2$ are linear operators. Therefore,

$$\hat{y}_i - y_i - (\hat{r}_i - r_i) = \beta \int_{t_k^i}^{t} \sum_{j=1}^{N} a_{ij}(\hat{w}_i - \hat{w}_j)d\tau + d(t_{k_i}^i) - d(t)$$
$$= \beta L_i^T \int_{t_k^i}^{t} z(\tau)d\tau + \Xi_i(z(t)) = \Phi_i(z(t)),$$

where $\Phi_i(\cdot) : \mathcal{L}_2 \mapsto \mathcal{L}_2$ is a linear operator.

Note also that

$$\int_{t_k^i}^{t} e^{-\theta(t-\tau)} \beta \sum_{j=1}^{N} a_{ij}(\hat{w}_i(\tau) - \hat{w}_j(\tau))d\tau = \beta L_i^T \int_{t_k^i}^{t} e^{-\theta(t-\tau)} z(\tau)d\tau$$
$$= \Gamma_i(z(t)),$$

where $\Gamma_i(\cdot) : \mathcal{L}_2 \mapsto \mathcal{L}_2$ is a linear operator. We then have

$$\epsilon_i(t) = \int_{t_k^i}^{t} e^{-\theta(t-\tau)} \theta \Phi_i(z(\tau))d\tau$$
$$+ \Gamma_i(z(t)) \int_{t_{k_i}^i}^{t} e^{-\theta(t-\tau)} \theta(\hat{r}_i(\tau) - r_i(\tau))d\tau$$
$$= \Psi_i(z(t)) + \int_{t_k^i}^{t} e^{-\theta(t-\tau)} \theta(\hat{r}_i(\tau) - r_i(\tau))d\tau,$$

where $\Psi_i(\cdot) : \mathcal{L}_2 \mapsto \mathcal{L}_2$ is a linear operator. Since r_i is bounded $\forall t > 0$, then we have $\epsilon_i(s) = \Delta_i^e(z(s))$ where $\Delta_i^e(\cdot) : \mathcal{L}_2 \mapsto \mathcal{L}_2$ is a nonlinear operator in the frequency domain. It then remains to determine the operator gain $\|\Delta^e\|_\infty$. This procedure is quite similar to the counterpart of Theorem 7.1 and is omitted here for brevity. □

Remark 7.9 Different from the static event-triggered onsensus problem in the previous sections, due to the introduction of the exogenous reference signals, the sampling errors in the dynamic average consensus problem are no longer images of linear operators acting on the consensus error of the sampled states. Nevertheless, the triggering function ensures that these operators are finite-gain $\mathcal{L}_2$ stable and thus can be handled together with the additive dynamic uncertainties using the operator-theoretic approach.

Define the dynamic average consensus tracking error of agent i as $e_i = x_i - \frac{1}{N}\sum_{i=1}^{N} r_i$. Let $p_i = x_i - r_i$, $p = [p_1, \cdots, p_N]^T$, $e = [e_1, \cdots, e_N]^T$, $d = [d_1, \cdots, d_N]^T$, $r = [r_1, \cdots, r_N]^T$, $u = [u_1, \cdots, u_N]^T$, and $w = [w_1, \cdots, w_N]^T$.

The dynamic average consensus system can then be rewritten in a compact form:

$$\begin{bmatrix} \dot{p} \\ \dot{w} \end{bmatrix} = \begin{bmatrix} 0 & -\beta L \\ \theta I & -\theta I - \beta L \end{bmatrix} \begin{bmatrix} p \\ w \end{bmatrix} + \begin{bmatrix} 0 & -\beta L & 0 \\ \theta I & -\beta L & \theta I \end{bmatrix} \begin{bmatrix} d \\ \epsilon \\ r \end{bmatrix},$$
$$\begin{bmatrix} u \\ z \\ e \end{bmatrix} = \begin{bmatrix} 0 & -\beta L \\ 0 & M \\ I & 0 \end{bmatrix} \begin{bmatrix} p \\ w \end{bmatrix} + \begin{bmatrix} 0 & -\beta L & 0 \\ 0 & M & 0 \\ 0 & 0 & M \end{bmatrix} \begin{bmatrix} d \\ \epsilon \\ r \end{bmatrix}. \tag{7.25}$$

Denoting $\tilde{u} = U^T u$, $\bar{u} = \tilde{u}_{2:N}$, $\tilde{z} = U^T z$, $\bar{z} = \tilde{z}_{2:N}$, $\tilde{d} = U^T d$, $\bar{d} = \tilde{d}_{2:N}$, $\tilde{\epsilon} = U^T \epsilon$, $\bar{\epsilon} = \tilde{\epsilon}_{2:N}$, $\tilde{p} = U^T p$, $\bar{p} = \tilde{p}_{2:N}$, $\tilde{w} = U^T w$, $\bar{w} = \tilde{w}_{2:N}$, $\tilde{e} = U^T e$, and $\tilde{r} = U^T r$, we have

$$\begin{bmatrix} \dot{\bar{p}} \\ \dot{\bar{w}} \end{bmatrix} = \begin{bmatrix} 0 & -\beta \bar{\Lambda} \\ \theta I & -\theta I - \beta \bar{\Lambda} \end{bmatrix} \begin{bmatrix} \bar{p} \\ \bar{w} \end{bmatrix} + \begin{bmatrix} 0 & -\beta \bar{\Lambda} & 0 \\ \theta I & -\beta \bar{\Lambda} & \theta I \end{bmatrix} \begin{bmatrix} \bar{d} \\ \bar{\epsilon} \\ \tilde{r} \end{bmatrix},$$
$$\begin{bmatrix} \bar{u} \\ \bar{z} \\ \tilde{e} \end{bmatrix} = \begin{bmatrix} 0 & -\beta \bar{\Lambda} \\ 0 & I \\ I & 0 \end{bmatrix} \begin{bmatrix} \bar{p} \\ \bar{w} \end{bmatrix} + \begin{bmatrix} 0 & -\beta \bar{\Lambda} & 0 \\ 0 & I & 0 \\ 0 & 0 & M \end{bmatrix} \begin{bmatrix} \bar{d} \\ \bar{\epsilon} \\ \tilde{r} \end{bmatrix}. \tag{7.26}$$

Note that

$$\bar{d} = Y^T \Delta^a (Y \bar{u}) = \bar{\Delta}^a (\bar{u}),$$
$$\bar{\epsilon} = Y^T \Delta^e (Y \bar{z}) = \bar{\Delta}^e (\bar{z}),$$

where $\bar{\Delta}^a, \bar{\Delta}^b : \mathcal{L}_2 \mapsto \mathcal{L}_2$ are nonlinear finite-gain operators with $\|\bar{\Delta}^a\|_\infty \leq \eta$ and $\|\bar{\Delta}^e\|_\infty \leq \sqrt{\alpha \lambda_N}$.

In this section, we aim to examine the robustness of the event-triggered dynamic average consensus algorithm (7.22) and (7.23) under additive dynamic uncertainties, which lies in two aspects. First, when there is no external reference signal r, the states x and w in the dynamic average consensus algorithm achieve robust consensus. Second, when there is a reference signal r, we want to see how small the H_∞ norm of the transfer function from the reference signal r to the average tracking error e, i.e., $\|T_{re}\|_\infty$, could be.

We now address the first problem and assume that the reference signal r does not exist temporarily. It is obvious that x and w reach the consensus if and only if $\bar{p}$ and $\bar{w}$ are both asymptotically stable. It is easy to derive from (7.26) that

$$\begin{bmatrix} \bar{d} \\ \bar{\epsilon} \end{bmatrix} = \begin{bmatrix} \bar{\Delta}^a & 0 \\ 0 & \bar{\Delta}^e \end{bmatrix} \begin{bmatrix} \bar{u} \\ \bar{z} \end{bmatrix},$$
$$\begin{bmatrix} \bar{u} \\ \bar{z} \end{bmatrix} = \begin{bmatrix} G_{11} & G_{12} \\ G_{21} & G_{22} \end{bmatrix} \begin{bmatrix} \bar{d} \\ \bar{\epsilon} \end{bmatrix}, \tag{7.27}$$

where

$$
\begin{aligned}
G_{11} &= -\theta\beta s\bar{\Lambda}\Pi,\ G_{12} = -\beta(s^2+s\theta)\bar{\Lambda}\Pi,\\
G_{21} &= s\theta\Pi,\ G_{22} = (s^2+s\theta)\Pi,\\
\Pi &= (s^2 I + s(\theta I + \beta\bar{\Lambda}) + \theta\beta\bar{\Lambda})^{-1}.
\end{aligned}
$$

Following similar lines in proving Theorem 7.4, we can obtain the following theorem.

Theorem 7.7 *Let* $\gamma = \max\{\eta, \sqrt{2\alpha\lambda_N}\}$. *The uncertain network system* (7.22) *reaches robust consensus under the event-triggered dynamic average consensus algorithm* (7.23), *if* $\frac{\theta\beta\lambda_N}{\theta+\beta\lambda_N} + 1 < \frac{1}{\gamma}$.

Remark 7.10 This theorem unveils the quantitative relationship among the robustness of the event-triggered dynamic average consensus algorithm against aperiodic event triggering and additive dynamic uncertainties, the parameters θ and β in the dynamic average consensus algorithm, and the largest eigenvalue of the Laplacian matrix.

Next, we move on to consider the dynamic average consensus tracking performance, quantified by $\|T_{re}\|_\infty$. Given the reference signal r, the smaller the $\|T_{re}\|_\infty$ is, the smaller the tracking error would be. From (7.26), we have

$$
\begin{aligned}
\begin{bmatrix}\bar{d}\\ \bar{\epsilon}\end{bmatrix} &= \begin{bmatrix}\bar{\Delta}^a & 0\\ 0 & \bar{\Delta}^e\end{bmatrix}\begin{bmatrix}\bar{u}\\ \bar{z}\end{bmatrix},\\
\begin{bmatrix}\bar{u}\\ \bar{z}\\ \tilde{e}\end{bmatrix} &= \begin{bmatrix}G_{11} & G_{12} & G_{13}\\ G_{21} & G_{22} & G_{23}\\ G_{31} & G_{32} & G_{33}\end{bmatrix}\begin{bmatrix}\bar{d}\\ \bar{\epsilon}\\ \tilde{r}\end{bmatrix},
\end{aligned}
\tag{7.28}
$$

where

$$
\begin{aligned}
G_{13} &= \begin{bmatrix}0_{N-1} & -\beta s\theta\bar{\Lambda}\Pi\end{bmatrix},\\
G_{23} &= \begin{bmatrix}0_{N-1} & s\theta\Pi\end{bmatrix},\\
G_{31} &= \begin{bmatrix}0_{N-1} & -\Pi\theta\beta\bar{\Lambda}\end{bmatrix}^T,\\
G_{32} &= \begin{bmatrix}0_{N-1} & -\Pi(s\beta\bar{\Lambda}+\theta\beta\bar{\Lambda})\end{bmatrix},\\
G_{33} &= \begin{bmatrix}M - \theta\beta\tilde{\Pi}\Lambda\end{bmatrix},\\
\tilde{\Pi} &= (s^2 I + s(\theta I + \beta\Lambda) + \theta\beta\Lambda)^{-1}.
\end{aligned}
$$

Note that (7.28) is an upper linear fractional transformation form and the closed-loop transfer function is $T_{\tilde{r}\tilde{e}} = \mathcal{F}_u(G, \Delta)$. Therefore, the performance index $\|T_{re}\|_\infty = \|T_{\tilde{r}\tilde{e}}\|_\infty = \|\mathcal{F}_u(G, \Delta)\|_\infty$.

Theorem 7.8 *Suppose that the parameter α, θ, β, μ and ν are chosen such that*

$$(\frac{\theta}{\beta\lambda_2}+1)^2+(\frac{\theta\beta\lambda_N}{\theta+\beta\lambda_N}+1)^2+2\frac{\theta\beta\lambda_N}{\theta+\beta\lambda_N}+\frac{2}{\beta\lambda_2}<\frac{1}{\gamma^2}, \tag{7.29}$$

where $\gamma = \max\{\eta, \sqrt{2\alpha\lambda_N}\}$. Then, the dynamic average consensus tracking performance $\|T_{re}\|_\infty \le \frac{1}{\gamma}$ is achieved by the event-triggered protocol (7.23). Moreover, the closed-loop system does not exhibit the Zeno behavior.

Proof In light of Lemma 2.11, to achieve the robust tracking performance $\|T_{re}\|_\infty < \frac{1}{\gamma}$ for all $\begin{bmatrix} \bar{\Delta}^a & \\ & \bar{\Delta}^b \end{bmatrix} \in \begin{bmatrix} \mathbf{\Delta}_1 & \\ & \mathbf{\Delta}_2 \end{bmatrix}$ and $\left\| \begin{bmatrix} \bar{\Delta}^a & \\ & \bar{\Delta}^b \end{bmatrix} \right\|_\infty \le \gamma$, it is enough to have

$$\mu_{\tilde{\mathbf{\Delta}}}\left(\begin{bmatrix} G_{11} & G_{12} & G_{13} \\ G_{21} & G_{22} & G_{23} \\ G_{31} & G_{32} & G_{33} \end{bmatrix}\right) < \frac{1}{\gamma},$$

where

$$\tilde{\mathbf{\Delta}} = \begin{bmatrix} \mathbf{\Delta}_1 & & \\ & \mathbf{\Delta}_2 & \\ & & \mathbf{\Delta}_f \end{bmatrix}.$$

It can be derived by some simple calculations that

$$\|G_{11}(j\omega)\|^2 = \frac{\omega^2\theta^2\beta^2\lambda_N^2}{(\omega^2+\theta^2)(\omega^2+\beta^2\lambda_N^2)},$$

$$\|G_{12}(j\omega)\| = \sqrt{\frac{\beta^2\lambda_N^2\omega^2}{\omega^2+\beta^2\lambda_N^2}},$$

$$\|G_{21}(j\omega)\| = \sqrt{\frac{\omega^2\theta^2}{(\omega^2+\theta^2)(\omega^2+\beta^2\lambda_2^2)}},$$

$$\|G_{22}(j\omega)\|^2 = \frac{\omega^2}{\omega^2+\beta^2\lambda_2^2},$$

$$\|G_{13}(j\omega)\| = \sqrt{\frac{\theta^2\beta^2\lambda_N^2\omega^2}{(\omega^2+\theta^2)(\omega^2+\beta^2\lambda_N^2)}} \le \frac{\theta\beta\lambda_N}{\theta+\beta\lambda_N},$$

$$\|G_{31}(j\omega)\| = \sqrt{\frac{\theta^2\beta^2\lambda_N^2}{(\omega^2+\theta^2)(\omega^2+\beta^2\lambda_N^2)}} \le 1,$$

$$\|G_{23}(j\omega)\| = \sqrt{\frac{\theta^2\omega^2}{(\omega^2+\theta^2)(\omega^2+\beta^2\lambda_2^2)}} \le 1,$$

$$\|G_{32}(j\omega)\| = \sqrt{\frac{\beta^2\lambda_N^2}{\omega^2+\beta^2\lambda_N^2}} \le 1,$$

$$\|G_{33}(j\omega)\|^2 = \max_{i=2,\cdots,N} \frac{\omega^4+(\theta+\beta\lambda_i)^2\omega^2}{(\omega^2+\theta^2)(\omega^2+\beta^2\lambda_i^2)} \le (\frac{\theta}{\beta\lambda_2}+1)^2.$$

In light of Lemma 2.10, we can derive that $\|T_{re}\|_\infty < \frac{1}{\gamma}$, if (7.29) holds. The Zeno behavior can be similarly excluded as in Theorem 7.2. □

Remark 7.11 Not surprisingly, we can easily see from (7.29) that a necessary condition of the robust performance specification is $\frac{\theta\beta\lambda_N}{\theta+\beta\lambda_N}+1 < \frac{1}{\gamma}$. This coincides with the fact that the robust performance specification requires that the robust consensus specification holds when there are no reference signals. Moreover, it can also be observed from (7.29) that to ensure that the event-triggered dynamic average consensus algorithm has a better robustness and tracking performance under additive uncertainties, the proportional gain θ should be chosen to be as small as possible. Moreover, the parameter α in the event-triggered mechanism should be relatively small. This is because when α increases, the operator gain will be larger, which will lead to larger sampling errors and thus a worse control performance.

7.5 Simulation Results

In this section, simulation examples will be provided to validate the effectiveness of the theoretical results. Consider a multi-agent system with six agents whose communication graph is shown in Fig. 7.3.

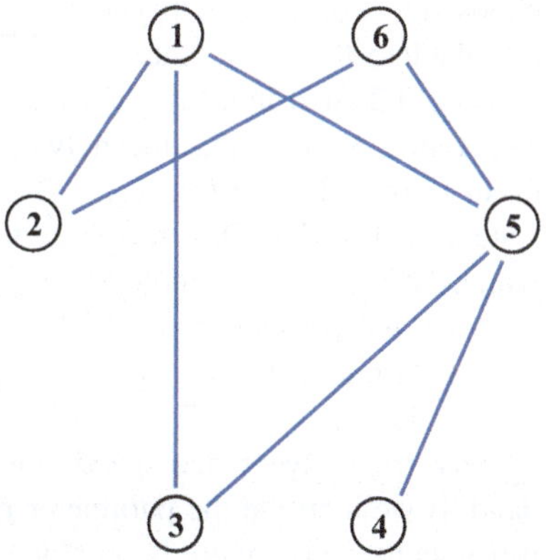

Fig. 7.3 The network graph

The Laplacian matrix of the graph is

$$L = \begin{bmatrix} 3.5 & -1 & -2 & 0 & -0.5 & 0 \\ -1 & 2 & 0 & 0 & 0 & -1 \\ -2 & 0 & 3 & 0 & -1 & 0 \\ 0 & 0 & 0 & 1 & -1 & 0 \\ -0.5 & 0 & -1 & -1 & 3.5 & -1 \\ 0 & -1 & 0 & 0 & -1 & 2 \end{bmatrix}.$$

We randomly generate transfer matrices representing the additive dynamic uncertainties from Δ_1^a to Δ_6^a as follows:

$$\Delta_1^a = \left[\begin{array}{cc|c} -55.4 & 140.7 & -8.24 \\ -155.7 & -71.41 & -1.28 \\ \hline 3.3989 & -5.4689 & 0.1022 \end{array}\right], \Delta_2^a = \left[\begin{array}{cc|c} -55.4 & 140.7 & -5.1520 \\ -163.7 & -39.56 & -9.2230 \\ \hline 2.03 & -2.14 & 0.2760 \end{array}\right],$$

$$\Delta_3^a = \left[\begin{array}{c|c} -0.3 & -0.28 \\ \hline 0.5454 & 0.0460 \end{array}\right], \Delta_4^a = \left[\begin{array}{cc|c} -5.4 & 14.7 & -1.24 \\ -15.7 & -1.41 & -0.28 \\ \hline 0.1150 & -2.4610 & 0.0920 \end{array}\right],$$

$$\Delta_5^a = \left[\begin{array}{cc|c} -55.4 & 140.7 & -3.0150 \\ -155.7 & -71.41 & -3.1122 \\ \hline 0.33 & -2.14 & 0.4133 \end{array}\right],$$

$$\Delta_6^a = \left[\begin{array}{ccc|c} -44.4 & 140.7 & -57.4 & -0.24 \\ -19.7 & -18.41 & -6.32 & -1.28 \\ 45.70 & 29 & -130.84 & 1.16 \\ \hline 4.4563 & -10.2542 & 4.2646 & 0.2875 \end{array}\right].$$

Note that $\Delta_i^a \in \mathcal{RH}_\infty$ $\forall i = 1, \cdots, 6$ and $\|\Delta_i^a\|_\infty \leq 0.4654 = \eta$. We then choose $\alpha = 0.02$, $\mu = 0.1$ and $\nu = 5$ in the triggering function (7.11). It is easy to calculate that $\gamma = \max\{\eta, \sqrt{2\alpha\lambda_6}\} = 0.4680$. We select the controller gain β to be 0.2 such that the robust consensus condition in Theorem 7.4 is satisfied. Theorem 7.4 states that the network reaches consensus asymptotically under the event-triggered algorithm (7.3). To illustrate this, we depict the evolution of the state x in Fig. 7.4. Denote the consensus error $\zeta = Mx$. The time instant after which $\|\zeta\| < 0.1$ is denoted as $t_{\min}$. When $\beta = 0.2$, $t_{\min} = 14.7304$ s. We then decrease the parameter β to 0.1 and let the other parameters remain unchanged. The evolution of the state x with respect to the time t is depicted in Fig. 7.5. When $\beta = 0.1$, $t_{\min} = 26.6796$ s, which is evidently larger than that when $\beta = 0.2$. Clearly, the convergence speed slows down with the decrease of β. On the other hand, if we increase the parameter β too much, then in virtue of Theorem 7.4, the robust consensus cannot be guaranteed. For example, if we increase β to 1.2, then

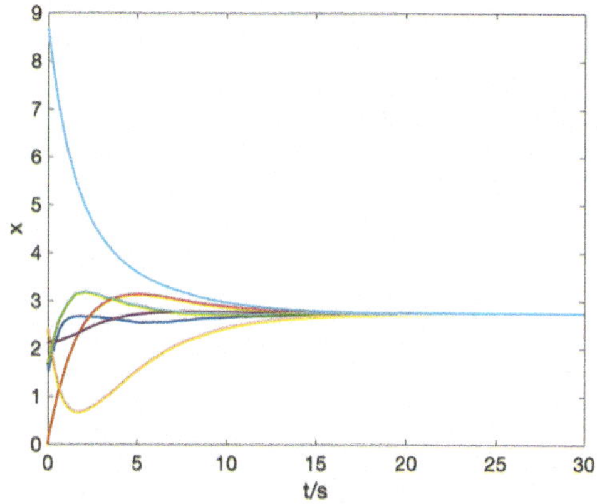

Fig. 7.4 The evolution of the state x when $\beta = 0.2$ under additive dynamic uncertainties

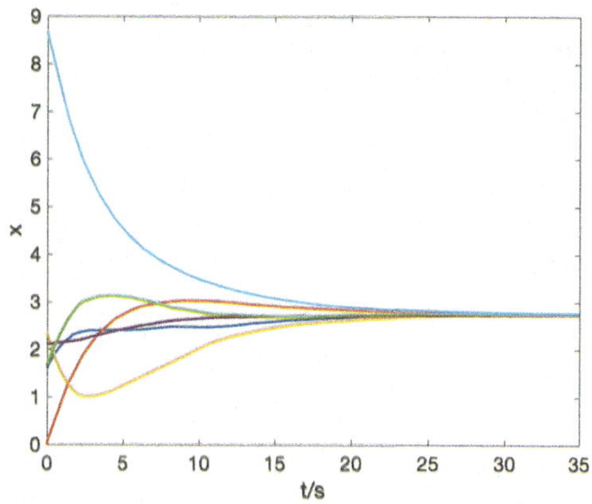

Fig. 7.5 The evolution of the state x when $\beta = 0.1$ under additive dynamic uncertainties

the state x cannot reach consensus anymore. To show that the closed-loop system does not exhibit the Zeno behavior, we draw the triggering instants of the agents when $\beta = 0.2$ in Fig. 7.6.

We next move on to test the robustness against network topology uncertainties and verify the result in Theorem 7.5. We consider the communication topology as in Fig. 7.3. For illustration, we randomly generate a dynamical perturbation for each edge within an H_∞ norm bound $\delta = 0.1315$, namely, Δ_i^c, $i = 1, \cdots, 7$. To satisfy the condition provided in Theorem 7.5, we choose $\beta = 0.08$, $\alpha = 0.002$, $\mu = 0.1$ and $\nu = 5$. It is easy to verify that $\gamma = \max\{\delta, \sqrt{2\alpha\lambda_6}\} = 0.1480$ and $\gamma < \sqrt{\frac{\lambda_2^2}{\lambda_2^2+\lambda_6^2}}$. The evolution of the states of the six agents is depicted in Fig. 7.7. It is shown that the six agents reach state consensus. The triggering instants of the agents are drawn in Fig. 7.8.

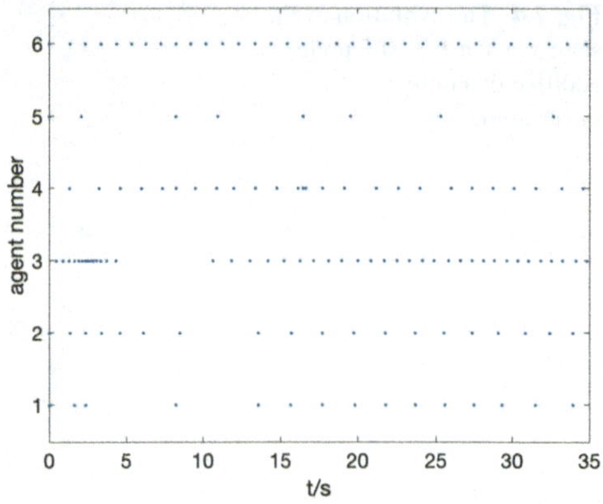

Fig. 7.6 Triggering instants of the six agents when $\beta = 0.2$ under additive dynamic uncertainties

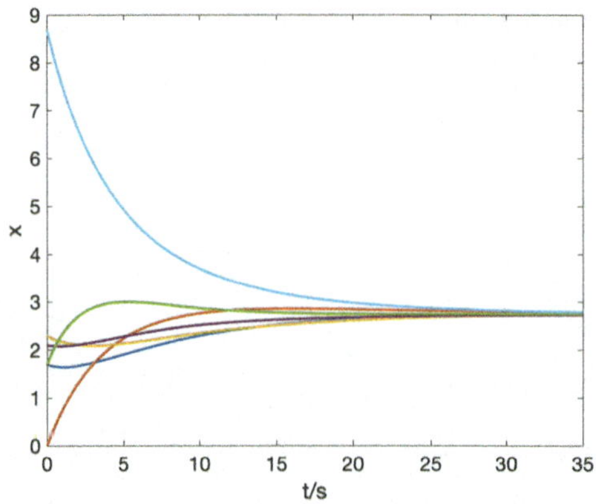

Fig. 7.7 The evolution of the state x under network topology uncertainties

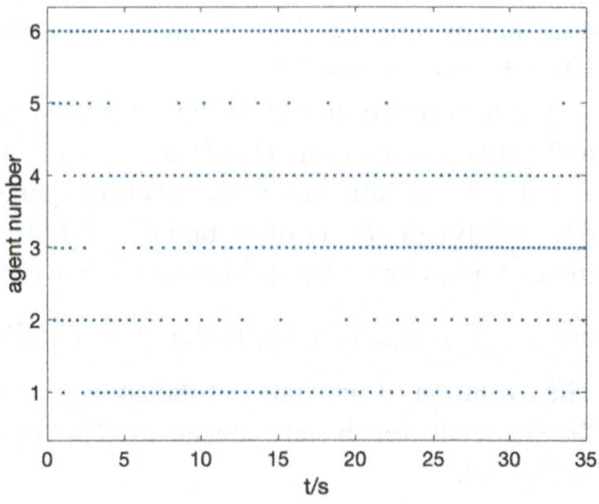

Fig. 7.8 Triggering instants of the six agents under network topology uncertainties

7.6 Notes

The materials of this chapter are mainly based on [115]. Note that the nominal agent dynamics are assumed to be single integrators here. The proposed operator-theoretic approach can be extended to high-order linear agent dynamics. The design of the robust event-triggered protocols nevertheless will be quite difficult. The robustness of event-triggered control of a single linear system with additive frequency-domain uncertainties is considered in [111].

Correction to: Distributed Event-triggered Control

Correction to:
Z. Li et al., *Distributed Event-triggered Control*,
https://doi.org/10.1007/978-981-99-8170-0

The book was inadvertently published with incorrect affiliation for Weihao Song as "Department of Control Science and Engineering, Tongji University, Shanghai, China." It has been updated as "Department of Mechanics and Engineering Science, Peking University, Beijing, China."

The updated version of this book can be found at
https://doi.org/10.1007/978-981-99-8170-0

Z. Li et al., *Distributed Event-triggered Control*,
https://doi.org/10.1007/978-981-99-8170-0_8

References

1. A. Adaldo, F. Alderisio, D. Liuzza, G.D. Shi, D.V. Dimarogonas, M. Bernardo, K.H. Johansson, Event-triggered pinning control of switching networks. IEEE Trans. Control Netw. Syst. **2**(2), 204–213 (2015)
2. K. Åström, B. Bernhardsson, Comparison of periodic and event based sampling for first-order stochastic systems, in *The 14th IFAC World Congress* (1999), pp. 301–306
3. J. Berneburg, C. Nowzari, Distributed dynamic event-triggered coordination with a designable minimum inter-event time, in *American Control Conference* (2019), pp. 1424–1429
4. D. Bernstein, *Matrix Mathematics: Theory, Facts, and Formulas* (Princeton University Press, Princeton, 2009)
5. Bitcraze, Crazyflie 2.1 (2023). https://www.bitcraze.io/products/crazyflie-2-1/
6. R. Cao, L. Cheng, Distributed dynamic event-triggered control for Euler-Lagrange multiagent systems with parametric uncertainties. IEEE Trans. Cybern. **53**(2), 1272–1284 (2023)
7. M. Cao, F. Xiao, L. Wang, Second-order consensus in time-delayed networks of integrators based on periodic edge-event driven control. Syst. Control Lett. **96**, 37–44 (2016)
8. Z. Chang, W. Song, J. Wang, Z. Li, Fully distributed event-triggered affine formation maneuver control over directed graphs. Sci. China Inf. Sci. **66**(6), 164201 (2023)
9. L. Chen, J. Mei, C. Li, G. Ma, Distributed leader-follower affine formation maneuver control for high-order multiagent systems. IEEE Trans. Automat. Control **65**(11), 4941–4948 (2020)
10. B. Cheng, B. He, Designing novel adaptive dynamic event-triggered protocols for uncertain multi-agent systems. Sci. China Inf. Sci. **67**(1), 119201 (2024)
11. B. Cheng, Z. Li, Consensus disturbance rejection with event-triggered communications. J. Franklin Inst. **356**(2), 956–974 (2019)
12. B. Cheng, Z. Li, Coordinated tracking control with asynchronous edge-based event-triggered communications. IEEE Trans. Automat. Control **64**(10), 4321–4328 (2019)
13. B. Cheng, Z. Li, Designing fully distributed adaptive event-triggered controllers for networked linear systems with matched uncertainties. IEEE Trans. Neural Netw. Learn. Syst. **30**(12), 3645–3655 (2019)
14. B. Cheng, Z. Li, Fully distributed event-triggered protocols for linear multi-agent networks. IEEE Trans. Automat. Control **64**(4), 1655–1662 (2019)
15. B. Cheng, Z. Li, Cooperative output regulation of heterogeneous multi-agent systems with adaptive edge-event-triggered strategies. IEEE Trans. Circuits Syst. II Express Briefs **67**(10), 2199–2203 (2020)
16. Y. Cheng, V. Ugrinovskii, Event-triggered leader-following tracking control for multivariable multi-agent systems. Automatica **70**, 204–210 (2016)

Z. Li et al., *Distributed Event-triggered Control*,
https://doi.org/10.1007/978-981-99-8170-0

17. T. Cheng, Z. Kan, J.R. Klotz, J.M. Shea, W.E. Dixon, Event-triggered control of multiagent systems for fixed and time-varying network topologies. IEEE Trans. Automat. Control **62**(10), 5365–5371 (2017)
18. B. Cheng, Y. Lv, Z. Li, Distributed adaptive event-triggered consensus with discrete control updating, in *The 59th IEEE Conference on Decision and Control* (2020), pp. 2793–2798
19. B. Cheng, Y. Lv, Z. Li, Z. Duan, Discrete communication and control updating in adaptive dynamic event-triggered consensus. IEEE Trans. Automat. Control (2023, in press). https://doi.org/10.1109/TAC.2023.3260681
20. C.A. Desoer, M. Vidyasagar, *Feedback Systems: Input-Output Properties* (SIAM, Philadelphia, 2009)
21. D.V. Dimarogonas, E. Frazzoli, K.H. Johansson, Distributed event-triggered control for multi-agent systems. IEEE Trans. Automat. Control **57**(5), 1291–1297 (2012)
22. X. Dong, G. Hu, Time-varying formation tracking for linear multiagent systems with multiple leaders. IEEE Trans. Automat. Control **62**(7), 3658–3664 (2017)
23. X. Fang, X. Li, L. Xie, Distributed formation maneuver control of multiagent systems over directed graphs. IEEE Trans. Cybern. **52**(8), 8201–8212 (2022)
24. M. Franceschelli, A. Gasparri, A. Giua, C. Seatzu, Decentralized estimation of laplacian eigenvalues in multi-agent systems. Automatica **49**(4), 1031–1036 (2013)
25. R.A. Freeman, P. Yang, K.M. Lynch, Stability and convergence properties of dynamic average consensus estimators, in *The 5th IEEE Conference on Decision and Control* (IEEE, 2006), pp. 338–343
26. M. Fu, L. Xie, The sector bound approach to quantized feedback control. IEEE Trans. Automat. Control **50**(11), 1698–1711 (2005)
27. E. Garcia, Y. Cao, H. Yu, A. Giua, P. Antsaklis, D. Casbeer, Decentralised event-triggered cooperative control with limited communication. Int. J. Control **86**(9), 1479–1488 (2013)
28. E. Garcia, Y. Cao, D.W. Casbeer, Decentralized event-triggered consensus with general linear dynamics. Automatica **50**, 2633–2640 (2014)
29. E. Garcia, Y. Cao, D.W. Casbeer, Periodic event-triggered synchronization of linear multi-agent systems with communication delays. IEEE Trans. Automat. Control **62**(1), 366–371 (2017)
30. B. Gharesifard, J. Cortés, Distributed strategies for generating weight-balanced and doubly stochastic digraphs. Eur. J. Control **18**(6), 539–557 (2012)
31. G. Guo, L. Ding, Q. Han, A distributed event-triggered transmission strategy for sampled-data consensus of multi-agent systems. Automatica **50**(5), 1489–1496 (2014)
32. W. He, S.S. Ge, Cooperative control of a nonuniform gantry crane with constrained tension. Automatica **66**, 146–154 (2016)
33. W. He, Y.H. Chen, Z. Yin, Adaptive neural network control of an uncertain robot with full-state constraints. IEEE Trans. Cybern. **46**(3), 620–629 (2016)
34. W.P.M.H. Heemels, J. Sandee, P. Van Den Bosch, Analysis of event-driven controllers for linear systems. Int. J. Control **81**, 571–590 (2008)
35. W.P.M.H. Heemels, K.H. Johansson, P. Tabuada, An introduction to event-triggered and self-triggered control, in *The 51st IEEE Conference on Decision and Control* (2012), pp. 3270–3285
36. W.P.M.H. Heemels, M.C.F. Donkers, A.R. Teel, Periodic event-triggered control for linear systems. IEEE Trans. Automat. Control **58**(4), 847–861 (2013)
37. T. Henningsson, E. Johannesson, A. Cervin, Sporadic event-based control of first-order linear stochastic systems. Automatica **44**(11), 2890–2895 (2008)
38. W. Hu, L. Liu, Cooperative output regulation of heterogeneous linear multi-agent systems by event-triggered control. IEEE Trans. Cybern. **47**(1), 105–116 (2017)
39. W. Hu, L. Liu, G. Feng, Output consensus of heterogeneous linear multi-agent systems by distributed event-triggered/self-triggered strategy. IEEE Trans. Cybern. **47**(8), 1914–1924 (2017)

40. W. Hu, L. Liu, G. Feng, Event-triggered cooperative output regulation of linear multi-agent systems under jointly connected topologies. IEEE Trans. Automat. Control **64**(3), 1317–1322 (2019)
41. W. Hu, Y. Weng, H. Li, Observer-based output feedback event-triggered robust H_∞ consensus control of uncertain linear multi-agent systems with directed networks. Int. J. Robust Nonlinear Control **32**(9), 5555–5573 (2022)
42. J. Huang, *Nonlinear Output Regulation: Theorem and Applications* (SIAM, Philadelphia, 2004)
43. N. Huang, Z. Sun, B.D.O. Anderson, Z. Duan, Distributed and adaptive triggering control for networked agents with linear dynamics. Inf. Sci. **517**, 297–314 (2020)
44. P. Ioannou, J. Sun, *Robust Adaptive Control* (Prentice-Hall, New York, 1996)
45. R.P. Jain, A.P. Aguiar, J. Sousa, Cooperative path following of robotic vehicles using an event-based control and communication strategy. IEEE Robotics Automat. Lett. **3**(3), 1941–1948 (2018)
46. H.J. Jongsma, H.L. Trentelman, M.K. Camlibel, Robust synchronization of Coprime factor perturbed networks. Syst. Control Lett. **95**, 62–69 (2016)
47. H.K. Khalil, *Nonlinear Systems* (Prentice-Hall, Englewood Cliffs, 2002)
48. S.S. Kia, J. Cortés, S. Martínez, Distributed event-triggered communication for dynamic average consensus in networked systems. Automatica **59**, 112–119 (2015)
49. S.S. Kia, B.V. Scoy, J. Cortes, R.A. Freeman, K.M. Lynch, S. Martinez, Tutorial on dynamic average consensus: The problem, its applications, and the algorithms. IEEE Control Syst. Mag. **39**(3), 40–72 (2019)
50. F.L. Lewis, H. Zhang, K. Hengster-Movric, A. Das, *Cooperative Control of Multi-Agent Systems: Optimal and Adaptive Design Approaches* (Springer-Verlag, London, 2014)
51. Z. Li, J. Chen, Robust consensus of linear feedback protocols over uncertain network graphs. IEEE Trans. Automat. Control **62**(8), 4251–4258 (2017)
52. Z. Li, . Chen, Robust consensus for multi-agent systems communicating over stochastic uncertain networks. SIAM J. Control Optim. **57**(5), 3553–3570 (2019)
53. Z. Li, Z. Duan, *Cooperative Control of Multi-agent Systems: A Consensus Region Approach* (CRC Press, Boca Raton, 2014)
54. Z Li, W. Ren, X. Liu, M. Fu, Consensus of multi-agent systems with general linear and Lipschitz nonlinear dynamics using distributed adaptive protocols. IEEE Trans. Automat. Control **58**(7), 1786–1791 (2013)
55. Z. Li, G. Wen, Z. Duan, W. Ren, Designing fully distributed consensus protocols for linear multi-agent systems with directed graphs. IEEE Trans. Automat. Control **60**(4), 1152–1157 (2015)
56. H. Li, G. Chen, T. Huang, Z. Dong, High-performance consensus control in networked systems with limited bandwidth communication and time-varying directed topologies. IEEE Trans. Neural Netw. Learn. Syst. **28**(5), 1043–1054 (2017)
57. X. Li, Y.C. Soh, L. Xie, Robust consensus of uncertain linear multi-agent systems via dynamic output feedback. Automatica **98**, 114–123 (2018)
58. X. Li, Y. Tang, H.R. Karimi, Consensus of multi-agent systems via fully distributed event-triggered control. Automatica **116**, 108898 (2020)
59. X. Li, Z. Sun, Y. Tang, H.R. Karimi, Adaptive event-triggered consensus of multi-agent systems on directed graphs. IEEE Trans. Automat. Control **66**(4), 1670–1685 (2021)
60. D. Li, G. Ma, Y. Xu, W. He, S.S. Ge, Layered affine formation control of networked uncertain systems: A fully distributed approach over directed graphs. IEEE Trans. Cybern. **51**(12), 6119–6130 (2021)
61. Z. Lin, L. Wang, Z. Chen, M. Fu, Z. Han, Necessary and sufficient graphical conditions for affine formation control. IEEE Trans. Automat. Control **61**(10), 2877–2891 (2016)
62. J. Liu, Y. Yu, Q. Wang, C.Y. Sun, Fixed-time event-triggered consensus control for multi-agent systems with nonlinear uncertainties. Neurocomputing **260**, 497–504 (2017)
63. Y. Lv, Z. Li, Is fully distributed adaptive protocol applicable to graphs containing directed spanning tree? Sci. China Inf. Sci. **65**(8), 189203 (2022)

64. Y. Lv, Z. Li, Z. Duan, J. Chen, Distributed adaptive output feedback consensus protocols for linear systems on directed graphs with a leader of bounded input. Automatica **74**, 308–314 (2016)
65. Y. Lv, Z. Li, Z. Duan, G. Feng, Novel distributed robust adaptive consensus protocols for linear multi-agent systems with directed graphs and external disturbances. Int. J. Control **90**(2), 137–147 (2017)
66. Y. Lv, G. Wen, T. Huang, Z. Duan, Adaptive attack-free protocol for consensus tracking with pure relative output information. Automatica **117**, 108998 (2020)
67. Y. Ma, J. Zhao, Distributed adaptive integral-type event-triggered cooperative output regulation of switched multiagent systems by agent-dependent switching with dwell time. Int. J. Robust Nonlinear Control **30**, 2550–2569 (2020)
68. J. Mei, W. Ren, J. Chen, Distributed consensus of second-order multi-agent systems with heterogeneous unknown inertias and control gains under a directed graph. IEEE Trans. Automat. Control **61**(8), 2019–2034 (2016)
69. X. Meng, T. Chen, Event based agreement protocols for multi-agent networks. Automatica **49**(7), 2125–2132 (2013)
70. X. Meng, L. Xie, Y.C. Soh, Asynchronous periodic event-triggered consensus for multi-agent systems. Automatica **84**, 214–220 (2017)
71. X. Meng, L. Xie, Y.C. Soh, Event-triggered output regulation of heterogeneous multi-agent networks. IEEE Trans. Automat. Control **63**(12), 4429–4434 (2018)
72. M. Mesbahi, M. Egerstedt, *Graph Theoretic Methods in Multiagent Networks* (Princeton University Press, Princeton, 2010)
73. C. Nowzari, J. Cortés, Distributed event-triggered coordination for average consensus on weight-balanced digraphs. Automatica **68**, 237–244 (2016)
74. C. Nowzari, E. Garcia, J. Cortés, Event-triggered communication and control of networked systems for multi-agent consensus. Automatica **105**, 1–27 (2019)
75. K.K. Oh, M.C. Park, H.S. Ahn, A survey of multi-agent formation control. Automatica **53**, 424–440 (2015)
76. R. Olfati-Saber, R. Murray, Consensus problems in networks of agents with switching topology and time-delays. IEEE Trans. Automat. Control **49**(9), 1520–1533 (2004)
77. R. Olfati-Saber, J. Fax, R. Murray, Consensus and cooperation in networked multi-agent systems. Proc. IEEE **95**(1), 215–233 (2007)
78. O. Onuoha, H. Tnunay, C. Wang, Z. Ding, Fully distributed affine formation control of general linear systems with uncertainty. J. Frank. Inst. **357**(17), 12143–12162 (2020)
79. OptiTrack, Prime^x 13 (2023). https://www.optitrack.com/cameras/primex-13/
80. Z. Peng, D. Wang, H. Zhang, G. Sun, Distributed neural network control for adaptive synchronization of uncertain dynamical multiagent systems. IEEE Trans. Neural Netw. Learn. Syst. **25**(8), 1508–1519 (2014)
81. Y. Qian, L. Liu, G. Feng, Output consensus of heterogeneous linear multi-agent systems with adaptive event-triggered control. IEEE Trans. Automat. Control **64**(6), 2606–2613 (2019)
82. Y. Qian, L. Liu, G. Feng, Distributed dynamic event-triggered control for cooperative output regulation of linear multiagent systems. IEEE Trans. Cybern. **50**(7), 3023–3032 (2020)
83. W. Ren, R.W. Beard, *Distributed Consensus in Multi-Vehicle Cooperative Control* (Springer-Verlag, London, 2008)
84. W. Ren, R.W. Beard, E.M. Atkins, Information consensus in multivehicle cooperative control. IEEE Control Syst. Mag. **27**(2), 71–82 (2007)
85. A. Rikos, T. Charalambous, C.N. Hadjicostis, Distributed weight balancing over digraphs. IEEE Trans. Control Netw. Syst. **1**(2), 190–201 (2014)
86. A. Seuret, C. Prieur, S. Tarbouriech, A.R. Teel, L. Zaccarian, A nonsmooth hybrid invariance principle applied to robust event-triggered design. IEEE Trans. Automat. Control **64**(5), 2061–2068 (2018)
87. G. Seyboth, D.V. Dimarogonas, K.H. Johanasson, Event-based broadcasting for multi-agent average consensus. Automatica **49**(1), 245–252 (2013)

88. H. Su, B. Cheng, Z. Li, Cooperative output regulation of heterogeneous systems over directed graphs: a dynamic adaptive event-triggered strategy. J. Syst. Sci. Complex. **36**, 909–921 (2022)
89. Z. Sun, N. Huang, B.D.O. Anderson, Z. Duan, Event-based multi-agent consensus control: Zeno-free triggering via $\mathcal{L}_p$ signals. IEEE Trans. Cybern. **50**(1), 284–296 (2020)
90. P. Tabuada, Event-triggered real-time scheduling of stabilizing control tasks. IEEE Trans. Automat. Control **52**(9), 1680–1685 (2007)
91. P. Tallapragada, N. Chopra, On event triggered tracking for nonlinear systems. IEEE Trans. Automat. Control **58**(9), 2343–2348 (2013)
92. H.L. Trentelman, K. Takaba, N. Monshizadeh, Robust synchronization of uncertain linear multi-agent systems. IEEE Trans. Automat. Control **58**(6), 1511–1523 (2013)
93. N.S. Tripathy, I. Kar, K. Paul, Stabilization of uncertain discrete-time linear system with limited communication. IEEE Trans. Automat. Control **62**(9), 4727–4733 (2016)
94. B. Wei, F. Xiao, M. Dai, Edge event-triggered control for multi-agent systems under directed communication topologies. Int. J. Control **91**(4), 887–896 (2018)
95. Y. Wu, R. Lu, Event-based control for network systems via integral quadratic constraints. IEEE Trans. Circuits Syst. I Reg. Pap. **65**(4), 1386–1394 (2018)
96. F. Xiao, X. Meng, T. Chen, Sampled-data consensus in switching networks of integrators based on edge events. Int. J. Control **88**(2), 391–402 (2015)
97. L. Xing, C. Wen, F. Guo, Z. Liu, H. Su, Event-based consensus for linear multiagent systems without continuous communication. IEEE Trans. Cybern. **47**(8), 2132–2142 (2017)
98. W. Xu, D.W.C. Ho, L. Li, J. Cao, Event-triggered schemes on leader-following consensus of general linear multiagent systems under different topologies. IEEE Trans. Cybern. **47**(1), 212–223 (2017)
99. Y. Xu, S. Zhao, D. Luo, Y. You, Affine formation maneuver control of high-order multi-agent systems over directed networks. Automatica **118**, 109004 (2020)
100. P. Yang, R.A. Freeman, G.J. Gordon, K.M. Lynch, S.S. Srinivasa, R. Sukthankar, Decentralized estimation and control of graph connectivity for mobile sensor networks. Automatica **46**(2), 390–396 (2010)
101. T. Yang, Z. Meng, D.V. Dimarogonas, K.H. Johansson, Global consensus for discrete-time multi-agent systems with input saturation constraints. Automatica **50**(2), 499–506 (2014)
102. D. Yang, W. Ren, X. Liu, W. Chen, Decentralized event-triggered consensus for linear multi-agent systems under general directed graphs. Automatica **69**, 242–249 (2016)
103. J. Yang, H. Yu, T. Chen, Affine formation maneuver control of event-triggered multi-agent systems. IFAC-PapersOnLine **53**(2), 3391–3396 (2020)
104. D. Yao, H. Li, R. Lu, Y. Shi, Distributed sliding-mode tracking control of second-order nonlinear multiagent systems: An event-triggered approach. IEEE Trans. Cybern. **50**(9), 3892–3902 (2020)
105. X. Yi, *Resource-Constrained Multi-Agent Control Systems: Dynamic Event-Triggering, Inputting Saturation, and Connectivity Preservation.* Licentiate Thesis, Rol Institute of Technology, Sweden, 2017
106. X. Yi, K. Liu, D.V. Dimarogonas, K.H. Johansson, Dynamic event-triggered and self-triggered control for multi-agent systems. IEEE Trans. Automat. Control **64**(8), 3300–3307 (2019)
107. P. Yu, L. Ding, Z. Liu, Z. Guan, Leader-follower flocking based on distributed event-triggered hybrid control. Int. J. Robust Nonlinear Control **26**, 143–153 (2016)
108. P. Yu, C. Fischione, D.V. Dimarogonas, Distributed event-triggered communication and control of linear multiagent systems under tactile communication. IEEE Trans. Automat. Control **63**(11), 3979–3985 (2018)
109. D. Zelazo, M. Bürger, On the robustness of uncertain consensus networks. IEEE Trans. Control Netw. Syst. **4**(2), 170–178 (2015)
110. J. Zhang, E. Fridman, Dynamic event-triggered control of networked stochastic systems with scheduling protocols. IEEE Trans. Automat. Control **66**(12), 6139–6147 (2021)

111. S. Zhang, Z. Li, Event-based robust stabilization: An operator-theoretic approach, in *the 61st IEEE Conference on Decision and Control* (IEEE, 2022), pp. 2373–2378
112. Z. Zhang, F. Hao, L. Zhang, L. Wang, Consensus of linear multi-agent systems via event-triggered control. Int. J. Control **87**(6), 1243–1251 (2014)
113. H. Zhang, G. Feng, H. Yan, Q. Chen, Observer-based output feedback event-triggered control for consensus of multi-agent systems. IEEE Trans. Ind. Electron. **61**(9), 4885–4894 (2014)
114. Y. Zhang, X. Su, Z. Liu, C.L.P. Chen, Event-triggered adaptive fuzzy tracking control with guaranteed transient performance for MIMO nonlinear uncertain systems. IEEE Trans. Cybern. **51**(2), 736–749 (2021)
115. S. Zhang, Y. Lv, Z. Li, An operator-theoretic approach to robust event-triggered control of network systems with frequency-domain uncertainties. IEEE Trans. Automat. Control **68**(4), 2034–2047 (2023)
116. S. Zhao, Affine formation maneuver control of multiagent systems. IEEE Trans. Automat. Control **63**(12), 4140–4155 (2018)
117. R. Zheng, X. Yi, W. Lu, T. Chen, Stability of analytic neural networks with event-triggered synaptic feedbacks. IEEE Trans. Neural Netw. Learn. Syst. **27**(2), 483–494 (2016)
118. K. Zhou, J.C. Doyle, *Essentials of Robust Control* (Prentice-Hall, Upper Saddle River, 1998)
119. W. Zhu, Z. Jiang, G. Feng, Event-based consensus of multi-agent systems with general linear models. Automatica **50**(2), 552–558 (2014)
120. W. Zhu, H. Li, Z. Jiang, Consensus of multi-agent systems with time-varying topology: an event-based dynamic feedback scheme. Int. J. Robust Nonlinear Control **27**(8), 1339–1350 (2016)
121. Z. Zuo, Q.-L. Han, B. Ning, *Fixed-Time Cooperative Control of Multi-Agent Systems* (Springer Nature, Switzerland AG, 2019)

GPSR Compliance

The European Union's (EU) General Product Safety Regulation (GPSR) is a set of rules that requires consumer products to be safe and our obligations to ensure this.

If you have any concerns about our products, you can contact us on ProductSafety@springernature.com

In case Publisher is established outside the EU, the EU authorized representative is:

Springer Nature Customer Service Center GmbH
Europaplatz 3
69115 Heidelberg, Germany

The manufacturer's authorised representative in the EU is Springer Nature Customer Service Centre GmbH, Europaplatz 3, 69115 Heidelberg, Germany. If you have any concerns regarding our products, please contact ProductSafety@springernature.com

Printed and bound by CPI Group (UK) Ltd, Croydon, CR0 4YY
30/09/2026
02225309-0015